AF898572

Respiratory Diseases Research at NIOSH

Reviews of Research Programs of the National Institute for Occupational Safety and Health

Committee to Review the NIOSH Respiratory Diseases Research Program

Board on Environmental Studies and Toxicology

Division on Earth and Life Studies

NATIONAL RESEARCH COUNCIL *AND*
INSTITUTE OF MEDICINE
OF THE NATIONAL ACADEMIES

THE NATIONAL ACADEMIES PRESS
Washington, D.C.
www.nap.edu

THE NATIONAL ACADEMIES PRESS • 500 Fifth Street, NW • Washington, DC 20001

NOTICE: The project that is the subject of this report was approved by the Governing Board of the National Research Council, whose members are drawn from the councils of the National Academy of Sciences, the National Academy of Engineering, and the Institute of Medicine. The members of the committee responsible for the report were chosen for their special competences and with regard for appropriate balance.

This project was supported by Contract No. 211-2006-19152 between the National Academy of Sciences and the National Institute for Occupational Safety and Health. Any opinions, findings, conclusions, or recommendations expressed in this publication are those of the author(s) and do not necessarily reflect the view of the organizations or agencies that provided support for this project.

International Standard Book Number-13: 978-0-309-11873-6
International Standard Book Number-10: 0-309-11873-5

Cover: Photograph by Keith Weller, U.S. Department of Agriculture. The air-curtain orchard sprayer uses multiple crossflow fans to disperse pesticide to apple trees.

Additional copies of this report are available from the National Academies Press, 500 Fifth Street, NW, Lockbox 285, Washington, DC 20055; (800) 624-6242 or (202) 334-3313 (in the Washington metropolitan area); Internet, *http://www.nap.edu.*

Printed in the United States of America.

Suggested citation: National Research Council and Institute of Medicine. 2008. *Respiratory Disesases Research at NIOSH.* Committee to Review the NIOSH Respiratory Diseases Research Program. Rpt. No. 4, Reviews of Research Programs of the National Institute for Occupational Safety and Health. Washington, D.C.: The National Academies Press.

THE NATIONAL ACADEMIES

Advisers to the Nation on Science, Engineering, and Medicine

The **National Academy of Sciences** is a private, nonprofit, self-perpetuating society of distinguished scholars engaged in scientific and engineering research, dedicated to the furtherance of science and technology and to their use for the general welfare. Upon the authority of the charter granted to it by the Congress in 1863, the Academy has a mandate that requires it to advise the federal government on scientific and technical matters. Dr. Ralph J. Cicerone is president of the National Academy of Sciences.

The **National Academy of Engineering** was established in 1964, under the charter of the National Academy of Sciences, as a parallel organization of outstanding engineers. It is autonomous in its administration and in the selection of its members, sharing with the National Academy of Sciences the responsibility for advising the federal government. The National Academy of Engineering also sponsors engineering programs aimed at meeting national needs, encourages education and research, and recognizes the superior achievements of engineers. Dr. Charles M. Vest is president of the National Academy of Engineering.

The **Institute of Medicine** was established in 1970 by the National Academy of Sciences to secure the services of eminent members of appropriate professions in the examination of policy matters pertaining to the health of the public. The Institute acts under the responsibility given to the National Academy of Sciences by its congressional charter to be an adviser to the federal government and, upon its own initiative, to identify issues of medical care, research, and education. Dr. Harvey V. Fineberg is president of the Institute of Medicine.

The **National Research Council** was organized by the National Academy of Sciences in 1916 to associate the broad community of science and technology with the Academy's purposes of furthering knowledge and advising the federal government. Functioning in accordance with general policies determined by the Academy, the Council has become the principal operating agency of both the National Academy of Sciences and the National Academy of Engineering in providing services to the government, the public, and the scientific and engineering communities. The Council is administered jointly by both Academies and the Institute of Medicine. Dr. Ralph J. Cicerone and Dr. Charles M. Vest are chair and vice chair, respectively, of the National Research Council.

www.national-academies.org

COMMITTEE TO REVIEW THE NIOSH RESPIRATORY DISEASES RESEARCH PROGRAM

Members

MARK UTELL (*Chair*), University of Rochester, New York
JOHN BALMES, University of California, San Francisco
PAUL BLANC, University of California, San Francisco
ELIZABETH CHAMBERLIN, Massey Coal Services, Chapmanville, WV
ROGENE HENDERSON, Lovelace Respiratory Research Institute, Albuquerque, NM
DAVID MANNINO, University of Kentucky, Lexington
JAMES MERCHANT, University of Iowa, Iowa City
JACQUELINE NOWELL, United Food and Commercial Workers International, Washington, DC
CHARLES POOLE, University of North Carolina, Chapel Hill
RICHARD SCHLESINGER, Pace University, New York, NY
NOAH SEIXAS, University of Washington, Seattle
IRA TAGER, University of California, Berkeley
DAVID WEGMAN, University of Massachusetts, Lowell

Staff

K. JOHN HOLMES, Senior Program Officer
KARL E. GUSTAVSON, Senior Program Officer
DAVID J. POLICANSKY, Scholar
RUTH E. CROSSGROVE, Senior Editor
CAY BUTLER, Editor
MIRSADA KARALIC-LONCAREVIC, Manager, Technical Information Center
JORDAN CRAGO, Senior Project Assistant
MORGAN R. MOTTO, Senior Program Assistant
RADIAH A. ROSE, Senior Editorial Assistant

Sponsor

NATIONAL INSTITUTE FOR OCCUPATIONAL SAFETY AND HEALTH

Preface

The National Institute for Occupational Safety and Health (NIOSH) was established by the Occupational Safety and Health Act of 1970 to "conduct . . . research, experiments, and demonstrations relating to occupational safety and health" and to develop "innovative methods, techniques, and approaches for dealing with [those] problems." One component of these activities is the Respiratory Diseases Research Program, whose stated mission is "to provide national and international leadership for the prevention of work-related respiratory diseases, using a scientific approach to gather and synthesize information, create knowledge, provide recommendations, and deliver products and services to those who can effect prevention." Work-related respiratory diseases are a serious problem of major magnitude. NIOSH indicates that deaths from work-related respiratory disease and malignancies account for about 70 percent of all occupational disease mortality.

In this report, the Committee to Review the NIOSH Respiratory Diseases Research Program evaluates NIOSH's Respiratory Diseases Research Program (RDRP). Using a framework developed by the NRC Committee to Review the NIOSH Research Program, this committee focused primarily on the last 10 years of RDRP activities to review the relevance and impact of RDRP's research portfolio. In its evaluation the committee also discusses and makes recommendations for upcoming research areas and challenges to be addressed by RDRP.

This report has been reviewed in draft form by persons chosen for their diverse perspectives and technical expertise in accordance with procedures approved by the NRC's Report Review Committee. The purpose of this independent review is to provide candid and critical comments that will assist the institution in making its published report as sound as possible and to ensure that the report meets institutional standards of objectivity, evidence, and responsiveness to the study charge. The review comments and draft manuscript remain confidential to protect the integrity of the deliberative process. We wish to thank the following for their review of this report: Lisa Brosseau, University of Minnesota; William Bunn III, International Truck & Engine Corporation; David Christiani, Harvard School of Public Health; Ellen Eisen, Harvard School of Public Health; Terry Gordon, New York University; Joel Kaufman, University of Washington; Franklin Mirer, Hunter College of the City University of New York; Roger McCarthy, Exponent, Inc.; Gunter Oberdorster, University of Rochester School of Medicine and Dentistry; Lorann Stallones, Colorado State University; and James Weeks, Advanced Technologies and Laboratories International, Inc.

Although the reviewers listed above have provided many constructive comments and suggestions, they were not asked to endorse the conclusions or recommendations, nor did they see the final draft of the report before its release. The review of this report was overseen by the review coordinator Frank Speizer, Harvard Medical School, and the review monitor, Mark Cullen, Yale University School of Medicine. Appointed by the NRC, they were responsible for making certain that an independent examination of this report was carried out in accordance with institutional procedures and that all review comments were carefully considered. Responsibility for the final content of this report rests entirely with the committee and the institution.

The committee gratefully acknowledges the following for making presentations to the committee: Michael Attfield, Vincent Castranova, Paul Henneberger, Kathleen Kreiss, Jacek Mazurek, Paul Schulte, Lewis Wade, David Weissman, and Ainsley Weston, National Institute for Occupational Safety and Health; Jim Cooper, Synthetic Organic Chemical Manufacturers Association; David Duebner, Brush Inc.; Amanda Edens and William Perry, Occupational Safety and Health Administration; Franklin E. Mirer, Hunter College; George Niewiadomski, Mine Safety and Health Administration; James Weeks, Advanced Technologies and Laboratories International, Inc. The committee also thanks Raymond Sinclair, National Institute for Occupational Safety and Health, for serving as the liaison to the committee and for his outstanding service and professionalism.

The committee is also grateful for the assistance of the NRC staff in preparing this report. Staff members who contributed to this effort are K. John Holmes, Karl Gustavson, and David Policansky, study directors; James Reisa, director of the

Board on Environmental Studies and Toxicology; Ruth Crossgrove and Cay Butler, editors; Mirsada Karalic-Loncarevic, research associate; and Morgan Motto, senior project assistant.

I would especially like to thank the committee for their efforts throughout the development of this report.

Mark J. Utell, *Chair*
Committee to Review the NIOSH
Respiratory Diseases Research Program

Contents

SUMMARY 1

1 INTRODUCTION 15
Overview of the Respiratory Diseases Research Program, 18
Study Charge and Evaluation Committee, 27
Evaluation Approach, 28
Organization of the Report, 32

2 EVALUATION OF THE RESPIRATORY DISEASES RESEARCH PROGRAM 34
RDRP Goals, Subgoals, and Resources, 35
External Factors Affecting the RDRP, 35
Surveillance, Health Hazard Evaluation and Technical Assistance Program, and Other Input Activities, 39
Introduction to Relevance and Impacts Assessment, 42
Strategic Goal 1: Prevent and Reduce Work-Related Airway Diseases, 43
Strategic Goal 2: Prevent and Reduce Interstitial Lung Diseases, 59
Strategic Goal 3: Prevent and Reduce Work-Related Infectious Respiratory Diseases, 81
Strategic Goal 4: Prevent and Reduce Work-Related Respiratory Malignancies, 97

Strategic Goal 5: Prevent Respiratory and Other Diseases Potentially Resulting from Occupational Exposures to Nanomaterials, 109
Overall Assessment of the RDRP Relevance and Impacts, 117

3 **EMERGING RESEARCH AND RESEARCH ISSUES** **120**
Strategic Goal 1: Prevent and Reduce Work-Related Airways Diseases, 120
Strategic Goal 2: Prevent and Reduce Work-Related Interstitial Lung Diseases, 122
Strategic Goal 3: Prevent and Reduce Work-Related Infectious Respiratory Diseases, 125
Strategic Goal 4: Prevent and Reduce Work-Related Respiratory Malignancies, 127
Strategic Goal 5: Prevent Respiratory and Other Diseases Potentially Resulting from Occupational Exposures to Nanomaterials, 128
Cross-Cutting Issues: Continuing and Emerging Issues Cutting Across Multiple Program Goals, 129

4 **RECOMMENDATIONS** **132**
Strategic Goal 1: Prevent and Reduce Work-Related Airway Diseases, 132
Strategic Goal 2: Prevent and Reduce Work-Related Interstitial Lung Diseases, 139
Strategic Goal 3: Prevent and Reduce Work-Related Infectious Respiratory Diseases, 141
Strategic Goal 4: Prevent and Reduce Work-Related Respiratory Malignancies, 143
Strategic Goal 5: Prevent Respiratory and Other Diseases Potentially Resulting from Occupational Exposures to Nanomaterials, 144
Cross-Cutting Issues, 145
Broader Programmatic Goals, 149

REFERENCES **150**

APPENDIXES
A Framework for the Review of Research Programs of the National Institute for Occupational Safety and Health 173
B Stakeholder Responses: Online Survey 217
C Bibliographic Information for the Committee to Review the NIOSH Respiratory Diseases Research Program 223
D Biographic Information on the Committee to Review the NIOSH Respiratory Diseases Research Program 226

Tables, Figures, and Boxes

TABLES

1-1 Timeline of Administrative Developments Related to NIOSH Respiratory Diseases Research, 19
1-2 NORA Priority Research Areas, 21
1-3 Example of NORA's Matrix Approach to Coordinating Research, 21
1-4 NORA2 Structure, 22
1-5 NIOSH RDRP Funding by Fiscal Year, 1996-2005, 25
1-6 NIH versus NIOSH Extramural Funding, 26

2-1 Strategic Goals and Subgoals of the NIOSH RDRP, as of February 2007, 36
2-2 NIOSH RDRP Budget and Staffing by Research Goals, 37
2-3 NIOSH Divisions, Laboratories, and Offices Involved in the RDRP, 39
2-4 Summary of Infectious Diseases Program Outputs and Outcomes, 82

A-1 NORA High-priority Research Areas by Category, 181
A-2 Examples of NIOSH Program Research and Transfer Activities, 198
A-3 Examples of a Variety of Scientific Information Outputs, 201
A-4 Evaluation Committee Worksheet to Assess Research Programs and Subprograms, 209

C-1 List of Documents and Other Items Submitted by the NIOSH RDRP to the National Academies Evaluation Committee between October 10, 2006 and May 17, 2007, 224

FIGURES

1-1 NIOSH organization chart, as of October 2006, 16
1-2 RDRP projects throughout NIOSH, 23
1-3 Flow chart for the evaluation of the NIOSH research program, 29
1-4 Logic model for the RDRP, 32

2-1 Proportional Allocation of RDRP Intramural Funds by Goal by Fiscal Year, 38
2-2 Reductions in respirable dust concentrations derived from MSHA compliance data, and prevalence of CWP among participants in the Coal Workers' Health Surveillance Program, 1970-2005, 67

A-1 The NIOSH operational plan presented as a logic model, 182
A-2 Flow chart for the evaluation of the NIOSH research program, 186

BOXES

S-1 Scoring Criteria for NIOSH Program Reviews from Framework Document, 5
S-2 Summary of Recommendations, 13

1-1 Criteria for Rating Relevance, 30
1-2 Criteria for Rating Impact, 31
1-3 Logic Model Terms and Examples, 33

2-1 Scoring Criteria for NIOSH Program Reviews from Framework Document, 118

4-1 Summary of Recommendations, 134

B-1 Letter with Questionnaire, 218

Abbreviations and Acronyms

AIHA	American Industrial Hygiene Association
ALFORD	Appalachian Laboratory for Occupational Respiratory Diseases
ALOSH	Appalachian Laboratory for Occupational Safety and Health
AMT	3-amino-5-mercapto-1,2,4,-triazole
AOEC	Association of Occupational and Environmental Clinics
APR	Air Purifying Respirators
ASTM	American Society for Testing and Materials
ATS	American Thoracic Society
ATSDR	Agency for Toxic Substances and Disease Registry
BDS	Biological Detection Systems
BEST	Board on Environmental Studies and Toxicology
BLS	Bureau of Labor Statistics
BRDPI	Biomedical Research and Development Price Index
BRFSS	Behavior Risk Factor Surveillance Survey
BSC	Board of Scientific Counselors
CBD	Chronic Beryllium Disease
CBR	Chemical, Biological, and Radiological
CBRN	Chemical, Biological, Radiological, and Nuclear
CDC	Centers for Disease Control and Prevention
COPD	Chronic Obstructive Pulmonary Disease

CPWR	Center to Protect Workers' Rights
CWP	Coal Workers' Pneumoconiosis
CSTE	Council of State and Territorial Epidemiologists
DART	Division of Applied Research and Technology
DBBS	Division of Biological and Behavioral Science
DELS	Division of on Earth and Life Sciences
DHHS	Department of Health and Human Services
DOD	Department of Defense
DOE	Department of Energy
DOI	Department of the Interior
DPSE	Division of Physical Sciences and Engineering
DRDS	Division of Respiratory Disease Studies
DSHEFS	Division of Surveillance, Hazard Evaluations and Field Studies
DSR	Division of Safety Research
EC	Evaluation Committee
EID	Education and Information Division
EPA	Environmental Protection Agency
FBI	Federal Bureau of Investigation
FC	Framework Committee
FDA	Food and Drug Administration
FDNY	Fire Department of New York
FEMA	Federal Emergency Management Agency
FEV_1	Forced Expiratory Volume in One Second
FTE	Full-Time Equivalent
FVC	Forced Vital Capacity
HELD	Health Effects Laboratory Division
HEPA	High-Efficiency Particulate Aerosol
HETA	HHE and Technical Assistance
HHE	Health Hazard Evaluation
HMW	High Molecular Weight
IARC	International Agency for Research on Cancer
ILO	International Labour Office
ILSI	International Life Sciences Institute
IOM	Institute of Medicine
ISO	International Standards Organization

JEM	Job-Exposure Matrix
LMW	Low Molecular Weight
MDI	Methylene Diisocyanate
MESA	Mine Enforcement and Safety Administration
MINER	Mine Improvement and New Emergency Response
MMWR	Morbidity and Mortality Weekly Report
MOU	Memorandum of Understanding
MSHA	Mine Safety and Health Administration
NA	National Academies
NAS	National Academy of Sciences
NACOSH	National Advisory Committee on Occupational Safety and Health
NCEH	National Center for Environmental Health
NCHS	National Center for Health Statistics
NCI	National Cancer Institute
NCID	National Center for Infectious Diseases
NFPA	National Fire Protection Association
NHANES	National Health and Nutrition Examination Survey
NHIS	National Health Interview Survey
NHLBI	National Heart, Lung, and Blood Institute
NIEHS	National Institute of Environmental Health Sciences
NIH	National Institutes of Health
NIOSH	National Institute for Occupational Safety and Health
NMAM	NIOSH Manual of Analytical Methods
NNI	National Nanotechnology Initiative
NOHSM	National Occupational Health Survey of Mining
NORA	National Occupational Research Agenda
NORMS	National Occupational Respiratory Mortality System
NPPTL	National Personal Protective Technology Laboratory
NRC	National Research Council
NSCWP	National Study of Coal Workers' Pneumoconiosis
NSET	Nanoscale Science, Engineering and Technology Subcommittee
NSSPM	National Surveillance System of Pneumoconiosis Mortality
NSTC	National Science and Technology Council
NTP	National Toxicology Program
NTRC	Nanotechnology Research Center
OHC	Office of Health Communications

OMB	Office of Management and Budget
ORA	Office of Regulatory Analysis
ORDS	Occupational Respiratory Disease Surveillance
OSHA	Occupational Safety and Health Administration
PAPR	Powered Air Purifying Respirator
PART	Program Assessment Rating Tool
PCR	Polymerase Chain Reaction
PEL	Permissible Exposure Limit
PHS	Public Health Service
PPE	Personal Protective Equipment
PRL	Pittsburgh Research Laboratory
RCF	Refractory Ceramic Fibers
RDRP	Respiratory Diseases Research Program
REL	Recommended Exposure Limit
SARS	Severe Acute Respiratory Syndrome
SENSOR	Sentinel Event Notification System for Occupational Risks
SPIROLA	Spirometry Longitudinal Data Analysis
SRL	Spokane Research Laboratory
TB	Tuberculosis
TC	Total Carbon
TDI	Toluene Diisocyanate
TIL	Total Inward Leakage
UMWA	United Mine Workers of America
USPS	United States Postal Service
UVGI	Ultraviolet Germicidal Irradiation
VA	Veterans' Administration
VOC	Volatile Organic Compound
WHO	World Health Organization
WoRLD	Work-Related Lung Disease
WRA	Work-Related Asthma

Summary

ABSTRACT *Respiratory diseases from occupational exposures have a major adverse effect on worker health, with the National Institute for Occupational Safety and Health (NIOSH) estimating that deaths from work-related respiratory disease and malignancies account for about 70% of all occupational disease mortality. NIOSH has a range of research activities in the Respiratory Diseases Research Program (RDRP) that are intended to protect worker health. These activities focus on airway diseases such as asthma and chronic obstructive pulmonary disease (COPD), interstitial lung diseases such as silicosis and asbestosis, respiratory infectious diseases including tuberculosis and avian influenza and severe acute respiratory syndrome (SARS), and respiratory cancers.*

This report assesses the relevance of the RDRP's activities to improve occupational safety and health and evaluates the impact that the program's research has had in reducing work-related hazardous exposures, illnesses, and injuries. It is one of multiple reviews based on a request from NIOSH to the National Academies to independently evaluate several NIOSH programs. The review by this committee was conducted on the basis of a framework established by a parent committee established by the National Academies. That framework document established a scoring system from 1 to 5 for impact and relevance, with 5 being the highest.

Relevance of the program was evaluated in terms of the degree of research priority and connection to improvements in workplace protection. Factors taken into account include the frequency and severity of health outcomes and the number of people at risk, the structure of the program, and the degree of consideration of stakeholders' input. The impact of the program's research is evaluated in terms of its contributions to worker health and safety. The committee has assigned a score of 5 in its rating of relevance. This reflects the judgment of the committee that the activities related to most of the strategic goals are in the highest-priority subject areas and highly relevant to improvements in workplace protection and that the RDRP is engaged in transfer activities at a significant level. The committee has assigned a score of 4 in its rating of impact. This reflects the committee's judgment that most subprograms within the RDRP have made major contributions to worker health and safety on the basis of end and well-accepted intermediate outcomes. If the committee had been given the option of providing non-integer scores, the score for program impact would have been between 4 and 5.

The committee provides recommendations intended to assist NIOSH in identifying opportunities to improve the relevance and impact of its research portfolio. These recommendations include an emphasis on the need to expand surveillance activities to detect exposure to those agents responsible for disease and to detect outbreaks of known diseases and identify new respiratory diseases affecting workers. The committee also recommends continued support of research on the development and improvement of methods to detect respiratory effects earlier and more accurately, continued research on the molecular mechanisms driving the pathogenesis of workplace respiratory diseases, the characterization of those agents (or attributes of those agents) responsible for adverse respiratory effects, evaluations of genetic variability that affects worker susceptibility, and research on improving respirator technology.

The committee's recommendations are important, despite the high scores given for relevance and impact. The high scores reflect the guidance for ranking established by the framework committee and the committee's recognition of the financial constraints the RDRP has operated under. The committee's evaluation is of necessity retrospective. The recommendations are prospective, and are meant to help ensure that the RDRP continues to maintain its progress toward its goal of protecting workers from respiratory disease.

BACKGROUND

NIOSH was established by the Occupational Safety and Health Act of 1970 to "conduct . . . research, experiments, and demonstrations relating to occupational safety and health" and to develop "innovative methods, techniques, and approaches for dealing with [those] problems" (Public Law 91-596, 84 STAT. 1590, 91st Congress, S.2193, December 29, 1970). As part of the Centers for Disease Control and Prevention within the Department of Health and Human Services, NIOSH is authorized to conduct research, training, and education related to worker health and safety; perform on-site investigations to evaluate hazards in the workplace;

recommend occupational health and safety standards, and fund research by other agencies and private organizations.

NIOSH has a wide portfolio of activities that support a number of industrial sectors including mining; health care and social attitude; manufacturing; transportation, warehousing, and utilities; construction; agriculture, forestry, and fishing; trade; and service. One component of these activities is the RDRP, whose stated mission is "to provide national and international leadership for the prevention of work-related respiratory diseases, using a scientific approach to gather and synthesize information, create knowledge, provide recommendations, and deliver products and services to those who can effect prevention." Work-related respiratory diseases are a serious problem of major magnitude. NIOSH indicates that deaths from work-related respiratory disease and malignancies account for about 70% of all mortality due to occupationally related disease.

RDRP interests and research spans all of NIOSH. The program represents a broad range of individuals and groups that do work relevant to occupational respiratory disease, including at several NIOSH divisions and laboratories. Preventing occupational respiratory disease has been a key part of the NIOSH research portfolio since the agency's inception in 1970 when, for example, health screening and research related to coal workers' pneumoconiosis (CWP) was undertaken per the Federal Coal Mine Health and Safety Act of 1969. Current work-related respiratory diseases and disorders that NIOSH is studying consist of a broad spectrum of adverse health effects including airway diseases such as asthma and COPD, interstitial lung diseases such as silicosis and asbestosis, respiratory infectious diseases such as tuberculosis and avian influenza and SARS, and respiratory cancers. These health effects can arise in a wide range of occupational settings and range from mild, reversible conditions to progressive fatal disorders and can be linked to short-term or long-term exposures. These health effects have tremendous implications for worker health and, by extension, the national economy.

CHARGE TO THE COMMITTEE

In September 2004, NIOSH requested that the National Academies review several NIOSH programs with respect to the impact and relevance of their work in reducing workplace injury and illness and to identify future directions that their work might take. The "Committee to Review NIOSH's Research Programs" (termed the "framework committee") was established as an oversight committee and created a framework document that will be used to evaluate individual programs. The RDRP was selected as one of the programs to undergo such a review. The National Research Council convened the Committee to Review the NIOSH Respiratory Diseases Research Program in late 2006. Briefly, the statement of task charges the committee with evaluating the program's impact and relevance to health and safety issues in the workplace and make recommendations for improvement. In conducting the review, the committee was tasked with assessing the relevance of the program's activities to the improvement of occupational safety and health and evaluating the impact that the program's research has had in reducing work-related hazardous exposures, illnesses, and injuries. The full statement of task is provided in Chapter 1. Per the assessment protocol the framework committee established, this committee was charged with rating the performance of the program for its relevance and impact on a scale of 1 to 5 (see Box S-1). The committee was also asked to provide a qualitative narrative assessment of the program's efforts and suggestions about emerging issues that the program should be prepared to address.

The evaluation committee was selected to include members with expertise in epidemiology, exposure assessment, industrial hygiene, inhalation toxicology, occupational medicine, and pulmonology. The committee reviewed the RDRP focusing on the period since 1996, recognizing that improvements in workers' health with regard to respiratory diseases that occurred during this period may be a result of earlier NIOSH research activities. The committee met three times between October 2006 and March 2007. The first two meetings were data-gathering sessions that included presentations by NIOSH staff and other invited speakers in open session where an opportunity was provided for stakeholders and the general public to comment. Site visits at the NIOSH Morgantown, Cincinnati, and Pittsburgh facilities were also held.

REVIEW OF THE RDRP

The review of the RDRP follows the organization of the program's strategic goals. Five goals, intended to support the RDRP mission statement provided above, were provided:

- Prevent and reduce work-related airway diseases.
- Prevent and reduce work-related interstitial lung diseases.
- Prevent and reduce work-related infectious respiratory diseases.
- Prevent and reduce work-related respiratory malignancies.
- Prevent respiratory and other diseases potentially resulting from occupational exposures to nanomaterials.

Each of these strategic goals has several subgoals and components on which NIOSH provided information to the committee. Several other RDRP activities cut across these subgoals—for example, surveillance, the health hazard evaluation and technical assistance program, emergency response and disaster preparedness, respirator policy, and sampling and analytical methods activities.

BOX S-1
Scoring Criteria for NIOSH Program Reviews from Framework Document

Rating of Impact

5 = Research program has made a major contribution to worker health and safety on the basis of end outcomes or well-accepted intermediate outcomes.
4 = Research program has made a moderate contribution on the basis of end outcomes or well-accepted intermediate outcomes; research program generated important new knowledge and is engaged in transfer activities, but well-accepted intermediate outcomes or end outcomes have not been documented.
3 = Research program activities or outputs are going on and are likely to improve worker health and safety (with an explanation of why not rated higher).
2 = Research program activities or outputs are going on and may result in new knowledge or technology, but only limited application is expected.
1 = Research activities and outputs are NOT likely to have any application.
NA = Impact cannot be assessed; program is not mature enough.

Rating of Relevance

5 = Research is in highest-priority subject areas and highly relevant to improvements in workplace protection; research results in, and NIOSH is engaged in, transfer activities at a significant level (highest rating).
4 = Research is in high-priority subject area and adequately connected to improvements in workplace protection; research results in, and NIOSH is engaged in, transfer activities.
3 = Research focuses on lesser priorities and is loosely or only indirectly connected to workplace protection; NIOSH is not significantly involved in transfer activities.
2 = Research program is not well integrated or well focused on priorities and is not clearly connected to workplace protection and is inadequately connected to transfer activities.
1 = Research in the research program is an ad hoc collection of projects, is not integrated into a program, and is not likely to improve workplace safety or health.

Assessment of Relevance and Impact

Chapter 2 of this report describes in detail the NIOSH activities, outputs, and outcomes relating to each of the RDRP's strategic goals and their subgoals; an assessment of the relevance and assessment of the impacts of those activities and outcomes follows that description. To develop scores for the program as a whole, the committee considered its assessment of the relevance and impacts of NIOSH activities directed at individual program subgoals. It then weighted them qualitatively to arrive at an overall program assessment. Relevance was evaluated in terms of the degree of research priority and connection to improvements in workplace protection. Factors taken into account include the frequency and severity of health outcomes and the number of people at risk, the structure of the program, and the degree of consideration of stakeholders' input. The impact of the program's research is evaluated in terms of its contributions to worker health and safety. The scoring criteria are described in Box S-1.

The committee has assigned a score of 5 in its rating of relevance. This reflects the judgment of the committee that the activities related to most of the subgoals are in the highest-priority subject areas and highly relevant to improvements in workplace protection and that the RDRP is engaged in transfer activities at a significant level. This is particularly true for activities related to interstitial lung diseases as well as many parts of the activities related to airways and infectious diseases and malignancies. The committee also noted that the activities related to nanotechnology were highly relevant, even though this emerging area has yet to see any impacts related to intermediate or end outcomes. Activities related to parts of some subprograms, including some of the activities related to malignancies and infectious diseases, do not reach this highest level of relevance as reflected in the assessment of the subprograms described in Chapter 2. But those NIOSH activities are in important research areas with some connection to improvements in workplace protection.

The committee has assigned a score of 4 in its rating of impact. This reflects the committee's judgment that most subprograms within the RDRP have made major contributions to worker health and safety on the basis of end and well-accepted intermediate outcomes, while others have had a smaller impact or the impact is not easily discernible. For example, the documented decrease in the prevalence of latex sensitization, early decreases (pre-2000) in CWP, and decreases in silicosis-related deaths in the mining industry have had a high impact on protecting worker health. In other cases—for example, the lack of new emerging infectious diseases or further spread of infectious diseases such as avian influenza and SARS—it is nearly impossible to disentangle the contribution NIOSH has had in limiting the onset or spread of respiratory diseases and resulting decrease in risk or death to exposed workers. However, insofar as the RDRP has played a major role in developing and disseminating personal and building respiratory protection systems and guidance documents and educational materials for their use, it is fair to credit the Infectious Diseases Program with helping to contain what could have been a larger disaster with greater cost in human lives. After much deliberation on how to weigh the assessments of different subprograms, the committee reached a consensus that the program was clearly better than that called for in a score of 4, but not in sum what the committee would rate a 5. If the committee had been given the option of providing non-integer scores, the score for program impact would have been between 4 and 5.

Identifying Emerging Issues and Research Areas

The committee assessed progress by the RDRP in targeting new research relevant to future improvements in workplace protection from occupational respiratory disease and sought to identify significant emerging research areas important to the mission of NIOSH. Several specific areas of interest relating to various subgoals are presented in Chapter 3. One issue that recurs throughout the discussion is the importance of robust surveillance for diseases and exposures to toxicants. This surveillance is necessary to determine the agents responsible for disease, to detect outbreaks of known diseases, and to identify new respiratory diseases affecting workers. Surveillance activities are critical to fulfilling the objectives of all the major program goals and to assessing the performance of current and future RDRP activities. The committee understands that the lack of financial and personnel resources and not the lack of awareness or expertise are the major causes of inadequate surveillance. While the needed resources are not likely to emerge in the near future, the committee strongly emphasizes this limitation as a continuing and emerging issue. Disease surveillance and exposure monitoring activities in the workplace typically require strong collaborations with industry, as seen in efforts on chronic beryllium disease, CWP and COPD, and respiratory malignancies. The committee supports ongoing, and encourages future, collaborations in studies of occupational diseases, particularly in the face of emerging evidence of a link between exposure and outcome.

The committee supports a range of ongoing RDRP activities that proactively address the incidence of future diseases and emerging diseases. These activities include the development and improvement of methods (e.g., spirometry, radiography, and novel biomarkers) to detect respiratory effects earlier and more accurately, continued research on the molecular mechanisms driving the pathogenesis of workplace-related respiratory diseases, characterization of those agents (or attributes of those agents) responsible for respiratory effects, and evaluations of genetic variability that affects worker susceptibility. A primary means for NIOSH to prevent diseases and protect worker health is by reducing and eliminating exposure to toxicants and pathogenic organisms. In this regard, NIOSH research on improving respirators is another essential means to address the threat of respiratory diseases among workers.

RECOMMENDATIONS FOR PROGRAM IMPROVEMENT

As indicated in the discussion about the scoring of the RDRP, the committee views past and current research efforts quite favorably. In this section, the committee provides summary recommendations and a brief justification for each of the program's strategic goals and for cross-cutting issues identified by the committee; see Box S-2 for a summary of these recommendations. The committee hopes these recommendations will help support NIOSH in identifying opportunities to improve the relevance and impact of its research portfolio. Chapter 4 of this report provides greater detail on these recommendations.

Strategic Goal 1: Prevent and Reduce Work-Related Airway Diseases

1. *Improve detection of work-related asthma, work-related fixed obstructive airway disease, and relevant exposures.* The committee is concerned that the new NORA2[1] industrial-sector-based priority-setting approach may lead to a decreased emphasis on needed disease-focused research related to airway diseases. The RDRP should systematically evaluate whether work-related asthma activities are being compromised under the new approach. Because the contribution of occupational exposures to the burden of adult asthma is high, work in pursuit of the four work-related asthma subgoals (preventing and reducing rubber-latex asthma and allergy among health care workers, preventing and reducing work-related asthma in the isocyanate-production industry, preventing and reducing work-related asthma related to nonindustrial indoor environmental quality, and improving detection of work-related asthma and relevant exposures, described in more detail in Chapter 4), can have a potentially great impact on improved occupational safety and health among the U.S. workforce. In terms of COPD, understanding the contribution of occupational exposures is difficult. To understand this issue, the committee strongly recommends that the RDRP continue to support population-based studies of associations between occupational exposures and COPD to better define groups of workers at greatest risk for planning preventive strategies. In the flavoring industry, the RDRP response to the initial identification of diacetyl-induced bronchiolitis obliterans has led to surveillance efforts in multiple locations in an effort to detect and prevent disease. The committee agrees that continued surveillance, prevention of exposures, and mechanistic research to better understand this disease should continue to be a high priority for the RDRP.

Strategic Goal 2: Prevent and Reduce Work-Related Interstitial Lung Diseases

2. *Continue and expand efforts to prevent coal workers' pneumoconiosis (CWP), silicosis, fiber-induced interstitial lung disease, and chronic beryllium disease.* The activities related to interstitial lung diseases form a critical core of the RDRP and

[1] The second phase of the National Occupational Research Agenda (NORA) process, which established a framework to guide and organize research activities in a fiscally constrained environment.

have provided well-documented improvements in occupational health. It is important that the RDRP continue to expand its activities in these areas so it can build on its earlier successes and respond to new challenges. In particular, the committee recommends giving high priority to research on the increasing incidence of CWP, "hot spots" of rapidly progressive CWP, and the possible role of concentration and duration of exposure, coal rank, and silica level in the rapidly progressive cases of CWP. Other important areas of research the committee recommends are experimental studies of silica-induced cytotoxicity and fibrogenesis and the development of control technologies that include silica substitutes, particle surface coatings, and dust reduction measures. In addition, fibers that are asbestiform, such as wincherite and tremolite, which were found as containments of vermiculite mined in Libby, Montana, or of synthetic origin, such as nylon flock and refractory ceramic fiber, require continued study with attention to fiber characteristics, such as cleavage fragments, and low-level exposures, respectively. The committee recommends that the RDRP target work in support of a new recommended OSHA standard that would lead to improved controls to reduce airborne and dermal exposure to beryllium in all workplaces where it is used. The committee recommends that the effectiveness of digital radiography in CWP surveillance should be an important continuing research priority, which will extend to all interstitial lung diseases.

Strategic Goal 3: Prevent and Reduce Work-Related Infectious Respiratory Diseases

3. *Continue to support efforts to protect workers from occupational exposures and to define mechanisms that make workers susceptible to respiratory infections. Enhance surveillance for outbreaks of known occupational respiratory infections as well as emerging respiratory infections. Develop an overarching structure for the infectious disease component and coordinate with other federal agencies to adopt technologies for the detection of bioterrorism agents for the protection of workers.* The RDRP's efforts on infectious diseases appropriately concentrate on preventing infection through the use of respirator controls and understanding the mechanisms that underlie susceptibility. More robust surveillance for disease outbreaks in occupational settings is needed. These three approaches represent NIOSH's primary tools to prevent and reduce known and unknown (emerging) respiratory infectious diseases. The committee recommends that the RDRP collect specific occupational TB surveillance data and explore ways to improve TB surveillance. It also should consider dropping its subgoal of preventing outbreaks of histoplasmosis, because no new RDRP research is planned and no resources are available for specific surveillance activities.

Strategic Goal 4: Prevent and Reduce Work-Related Respiratory Malignancies

4. *Develop a comprehensive plan for addressing respiratory malignancies in the workplace while assuring the integration of this plan with NIOSH and other federal agency research program efforts to study malignancies. Refocus research on diagnostic tools to research on biomarkers of exposure or early detection of risk specific to occupational cohorts.* The impact of the respiratory malignancies program has been strong with regard to the three specific carcinogenic exposures listed as subgoals by the RDRP: hexavalent chromium, silica, and diesel exhaust. Ongoing research on respiratory malignancies resulting from workplace exposures continues to address challenging problems related to occupational risk of lung cancer, and the RDRP has been effective in engaging stakeholders from industry and the workforce. To enhance their efforts, the RDRP needs to develop, in collaboration with other relevant federal agencies (National Cancer Institute, National Institute of Environmental Health Sciences, Food and Drug Administration, Department of Defense, Department of Labor), a coordinated planning process to address occupational cancer risks to provide a comprehensive approach to detection, surveillance, and prevention. The committee recommends that the RDRP ensure that respiratory malignancies are well integrated into an overall program of occupational cancer research, and not arbitrarily separated from those efforts. The RDRP should consider refocusing its research on biomarkers for early detection to biomarkers of exposure or to early detection that addresses the needs of specific workers at high occupational risk of contracting lung cancer. The RDRP should continue efforts to develop and validate exposure methods for diesel particulate matter, especially in the presence of other sources of carbon aerosols, and to validate the methods used to measure diesel particulate matter in coal mines. The RDRP should consider developing long-term follow-up of workers exposed to asbestos and should consider enhancing surveillance for asbestos-related risks.

Strategic Goal 5: Prevent Respiratory and Other Diseases Potentially Resulting from Occupational Exposures to Nanomaterials

5. *NIOSH should continue to play a leading role in informing and guiding national and international efforts to address potential occupational hazards and risks associated with the use of manufactured nanomaterials.* The growing recognition of the usefulness of nanomaterials in various industrial applications has created an urgent need to study the potential health effects of exposures to nanoparticles and methods to control exposures to these particles. The RDRP has taken a lead at the national and international levels to address these questions. The RDRP is well suited to continue to develop exposure assessment methods and technology

to monitor effective control of exposures to nanomaterials in work settings. The committee generally supports the RDRP's research efforts on nanomaterial toxicity, exposure, and dose-response as part of a coordinated effort with other federal agencies and with appropriate prioritization for resource allocation to this problem. However, the committee is concerned that available data—especially on human-health effects—might not be sufficient for quantitative risk assessments and therefore suggests that the RDRP should consider other approaches for dealing with the potential health impacts of these new materials in a precautionary manner.

Cross-Cutting Issues

Systems for Surveillance

6. *NIOSH should provide appropriate resources for and engage in high-priority occupational disease surveillance.* The United States is practically alone among highly developed countries with regard to its lack of comprehensive surveillance of occupational diseases. The effectiveness of past NIOSH surveillance activities for coal-dust-related diseases, both CWP and COPD, highlight the importance of improved surveillance for other occupational respiratory disorders. NIOSH should engage in development and evaluation of surveillance methods as a high priority for occupational respiratory disease surveillance, including methods to systematically review and analyze the findings of reports through the health hazard evaluation and technical assistance program and other data collection approaches.

Exposure Assessment

7. *Produce a programmatic approach to the development of sampling and analytic methods that include exposure assessment scientists as an integral part of RDRP activities.* Exposure assessment is a core component of occupational respiratory disease research and prevention activities. However, the RDRP does not present an explicit or comprehensive focus on exposure assessment methods. Also, while the RDRP has focused on exposure assessment research in the past, no specific mention of current or future needs for exposure assessment activities is explicitly made.

Emergency Response

8. *The RDRP is encouraged to explore research strategies in its emergency response efforts. For example, the RDRP should assess how NIOSH-supported research and medical surveillance of World Trade Center disaster emergency responders and recovery workers may or may not be relevant to work-related asthma, work-related*

fixed obstructive airway disease, interstitial lung diseases, and possibly malignancies. The RDRP has made important contributions to the emergency response to recent disasters, including the World Trade Center and anthrax terror attacks, and hurricanes Katrina and Rita. However, much more can be learned about exposure-response relationships and ultimately about protecting emergency responders by conducting longitudinal cohort studies relating to catastrophic events. Information from emergency responses to toxicant exposures should be applicable to models of irritant-induced asthma, fixed airway obstruction, interstitial lung disease, and possibly even malignancies. The RDRP is encouraged to continue to develop cooperative work with other agencies that have mandates in infection and terrorism.

RDRP Resource Allocation

9. *The RDRP should prioritize all research proposals under consideration for funding, whether intramural or extramural, according to the RDRP strategic plan, which needs to be updated periodically.* The RDRP has recently been organized to emphasize sector-based as opposed to disease-based research; an emerging issue is how research priorities for respiratory diseases that cut across sectors will be treated, particularly as the RDRP encompasses many divisions and laboratories across NIOSH. The RDRP needs systems to govern the awarding of extramural grants, contracts, and cooperative agreements and to integrate the results of this external research into the intramural program. This system should ensure that unnecessary duplication and inappropriate expenditure on low-priority research projects are avoided.

Importance of the Committee's Recommendations

The committee's recommendations are important, despite the high scores given for relevance and impact. The high scores reflect the guidance for ranking established by the framework committee and the committee's recognition of the financial constraints the RDRP has operated under. The committee's evaluation is of necessity retrospective. The recommendations are prospective, and are meant to help ensure that the RDRP continues to maintain its progress toward its goal of protecting workers from respiratory disease.

BOX S-2
Summary of Recommendations

Strategic Goal 1: Prevent and Reduce Work-Related Airway Diseases
1. Improve detection of work-related asthma, work-related fixed obstructive airway disease, and relevant exposures.

Strategic Goal 2: Prevent and Reduce Work-Related Interstitial Lung Diseases
2. Continue and expand efforts to prevent coal workers' pneumoconiosis, silicosis, fiber-induced interstitial lung disease, and chronic beryllium disease.

Strategic Goal 3: Prevent and Reduce Work-Related Infectious Respiratory Diseases
3. Continue to support efforts to protect workers from occupational exposures and to define mechanisms that make workers susceptible to respiratory infections. Enhance surveillance for outbreaks of known occupational respiratory infections as well as emerging respiratory infections. Develop an overarching structure for the infectious disease component and coordinate with other federal agencies to adopt technologies for the detection of bioterrorism agents for the protection of workers.

Strategic Goal 4: Prevent and Reduce Work-Related Respiratory Malignancies
4. Develop a comprehensive plan for addressing respiratory malignancies in the workplace while assuring the integration of this plan with NIOSH and other federal agency research program efforts to study malignancies. Refocus research on diagnostic tools to research on biomarkers of exposure or early detection of risk specific to occupational cohorts.

Strategic Goal 5: Prevent Respiratory and Other Diseases Potentially Resulting from Occupational Exposures to Nanomaterials
5. NIOSH should continue to play a leading role in informing and guiding national and international efforts to address potential occupational hazards and risks associated with the use of manufactured nanomaterials.

Cross-Cutting Issues

Systems for Surveillance
6. NIOSH should provide appropriate resources for and engage in high-priority occupational disease surveillance.

Exposure Assessment
7. Produce a programmatic approach to the development of sampling and analytic methods that include exposure assessment scientists as an integral part of RDRP activities.

continued

BOX S-2 Continued

Emergency Response

8. The RDRP is encouraged to explore research strategies in its emergency response efforts. For example, the RDRP should assess how NIOSH-supported research and medical surveillance of World Trade Center disaster emergency responders and recovery workers may or may not be relevant to work-related asthma, work-related fixed obstructive airway disease, interstitial lung diseases, and possibly malignancies.

RDRP Resource Allocation

9. The RDRP should prioritize all research proposals under consideration for funding, whether intramural or extramural, according to the RDRP strategic plan, which needs to be updated periodically.

1

Introduction

The National Institute for Occupational Safety and Health (NIOSH) was established by the Occupational Safety and Health Act of 1970 to "conduct . . . research, experiments, and demonstrations relating to occupational safety and health" and to develop "innovative methods, techniques, and approaches for dealing with [those] problems" (Public Law 91-596, 84 STAT. 1590, 91st Congress, S.2193, December 29, 1970). NIOSH, part of the Centers for Disease Control and Prevention within the Department of Health and Human Services, is authorized to conduct research, training, and education related to worker health and safety; perform on-site investigations to investigate hazards in the workplace; recommend occupational health and safety standards; and fund research by other agencies or private organizations. NIOSH has the legislative responsibility to develop the research base upon which it can then recommend occupational health and safety standards to OSHA, although it does not have the authority to promulgate binding standards or enforce regulations on workplace safety and health. Federal regulatory and enforcement authority for occupational safety and health rests with the Occupational Safety and Health Administration (OSHA) and the Mine Safety and Health Administration (NIOSH 2006a). The organizational configuration of NIOSH is shown in Figure 1-1.

Preventing occupational respiratory disease has been a key part of the NIOSH research portfolio since the agency's inception in 1970. Early on, NIOSH assumed responsibilities for health screening and research related to coal workers' pneumoconiosis (CWP) that initially was mandated under the Federal Coal Mine Health and Safety Act of 1969. Respiratory disease research and surveillance related to min-

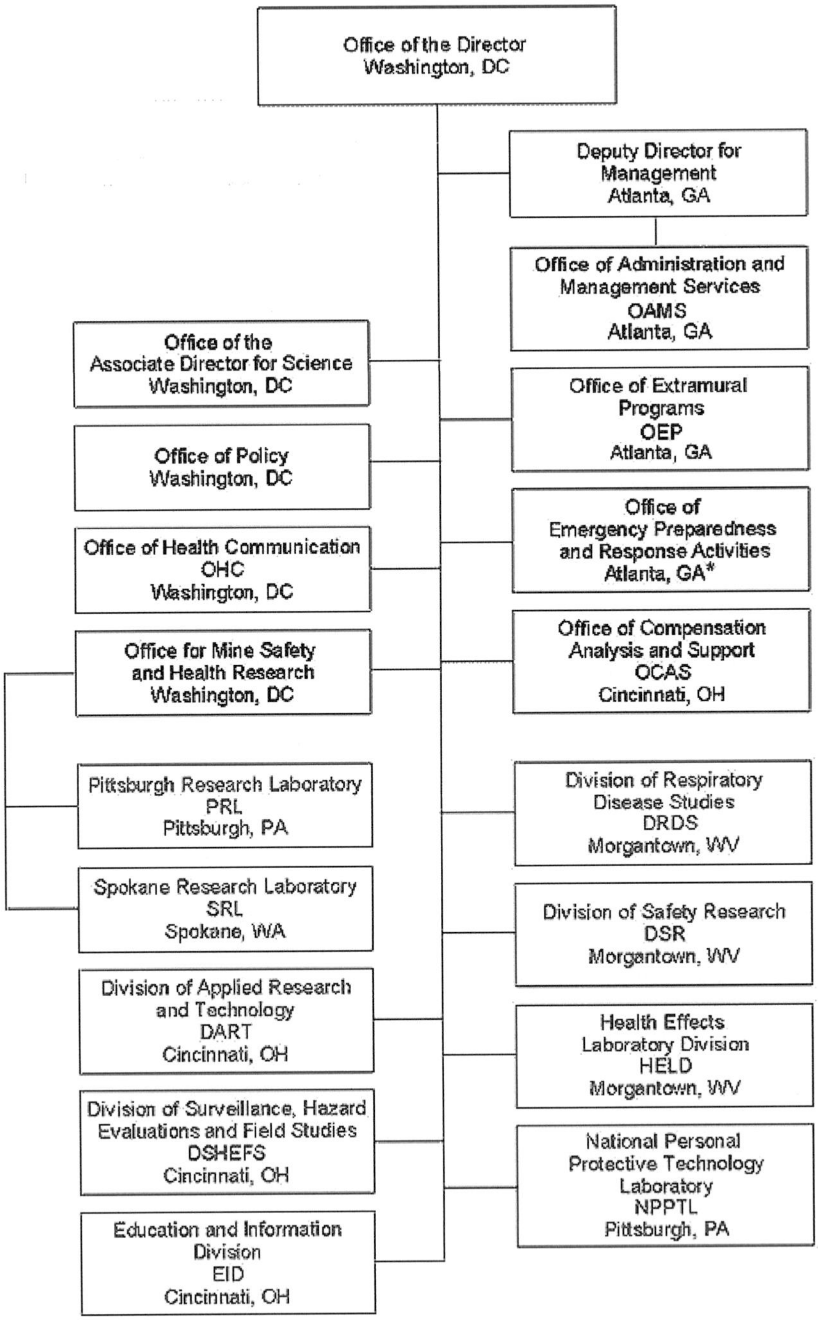

FIGURE 1-1 NIOSH organization chart, as of October 2006. Source: NIOSH 2006a.

ing activities remain major NIOSH activities. However, current work-related respiratory diseases and disorders that NIOSH is studying include a broad spectrum of other adverse health effects that can arise in a wide range of occupational settings. Such adverse health effects can range from mild, reversible conditions to progressive, fatal disorders and can be linked to short-term or long-term exposures.

A major focus of NIOSH's activities both historically and currently has been on occupational respiratory diseases. Deaths from work-related respiratory diseases and malignancies have been estimated to account for about 70% of all occupational disease mortality (Steenland et al. 2003). The American Thoracic Society stated that "A careful review of the literature demonstrates that approximately 15% of both asthma and COPD is likely to be work related" (Balmes et al. 2003). It has been estimated that this occupational asthma and chronic obstructive pulmonary disease (COPD) costs nearly $7 billion annually in the United States (Leigh et al. 2002). In 2000, there were an estimated 386,000 deaths worldwide from asthma, COPD, and pneumoconiosis and nearly 6.6 million disability-adjusted life years due to occupational exposure to airborne particulates (Driscoll et al. 2005a).

As work-related diseases have been identified and safety measures taken to reduce the associated risks, the overall patterns of occupational respiratory diseases have changed. NIOSH data through 1999 indicate that morbidity and mortality from CWP declined appreciably over the preceding 20 years (NIOSH 2003). More recent data, however, indicate that the trend has slowed and even reversed, making this area one of active NIOSH investigation (NIOSH 2006a; data from E. L. Petsonk, NIOSH, as cited by Ward 2007). In the same time period, mortality due to silicosis has declined from well over 1,000 deaths annually in the late 1960s to fewer than 200 per year in the late 1990s (NIOSH 2003). Nonetheless, as with CWP "hot spots" (MMWR 2006a, 2007a), NIOSH has continued to report important clusters of silica-related disease (MMWR 1990, 2004).

The identification of work-related outbreaks of respiratory disease—such as those related to artificial butter flavorings, respirable particles of nylon flock, and leather conditioning sprays—and of CWP hot spots point to the changing nature of occupational respiratory disease, with new issues arising from classic exposures as well as new diseases being detected from novel exposures. Additional challenges for NIOSH include the need to protect workers from potential occupational lung diseases due to exposures during national security emergencies, the potential for occupational exposures to weaponized biological agents, the emergence of naturally occurring infectious diseases, and the increased production and incorporation of nanomaterials that pose unknown risks to workers.

OVERVIEW OF THE RESPIRATORY DISEASES RESEARCH PROGRAM

The formally defined Respiratory Diseases Research Program (RDRP) is a recent creation. As described in the evidence package, the RDRP is an "organizational component" that was designated in 2005 as a result of matrix management efforts intended to coordinate cross-institute programmatic activities as a result of the second National Occupational Research Agenda (NORA2) process. NIOSH describes the RDRP as "the range of individuals and groups supported by NIOSH to do work that is relevant to occupational respiratory disease" (NIOSH 2006a). As such, the program is intended to be a vehicle to describe the broad range of research on respiratory disease that occurs at NIOSH. For example, the RDRP is not an institute or physical location where the research occurs nor does it constitute a cohesive staff grouping under a single hierarchical reporting structure. The RDRP includes the multiple divisions and laboratories within NIOSH that deal with respiratory disease issues. For the sake of clarity, throughout this report the committee uses the term RDRP to refer to activities that occurred before as well as after establishment of the RDRP in 2005, even though it is recognized that the designation, as such, did not exist in the earlier time frame.

History of the Program

Originating Legislation and Facilities

Table 1-1 shows some of the administrative developments important for development of the RDRP. The Federal Coal Mine Health and Safety Act of 1969 prescribed a number of research, surveillance, and regulatory-related activities that were later assigned to NIOSH when it was formed by the Occupational Health and Safety Act of 1970. Adoption of the Mine Safety and Health Act of 1977 also had an impact on the RDRP mission. Because of the location of NIOSH coal mining research facilities—first in Beckley, West Virginia, and later in Morgantown, West Virginia—initial RDRP research focused almost entirely on CWP. Beginning in 1976, with the formation of the Division of Respiratory Disease Studies (DRDS), the scope of research on respiratory disease based in Morgantown expanded significantly to other forms of pneumoconiosis that included silicosis, beryllium disease, asbestosis, and other fibrotic lung diseases, and a focus on organic dusts and other respiratory irritants that result in occupational asthma, airway obstruction, and hypersensitivity pneumonitis.

Other elements of NIOSH that have had a role in respiratory disease activities are located in NIOSH facilities in Cincinnati, Ohio. These facilities include the Division of Applied Research and Technology, with expertise in exposure assessment, development of analytical methods, and control technologies; the Education

TABLE 1-1 Timeline of Administrative Developments Related to NIOSH Respiratory Diseases Research

Year	Administrative Development
1969	Federal Coal Mine Health and Safety Act
1970	Occupational Health and Safety Act
	Several RDRP components established in Cincinnati, Ohio.
1971	Appalachian Laboratory for Occupational Safety and Health (ALOSH) opens in Morgantown, West Virginia
1976	Division of Respiratory Disease Studies opens within ALOSH
1977	Federal Mine Safety and Health Act of 1977
1996	Health Effects Laboratory Division founded
1996	Pittsburgh Research Laboratory and Spokane Research Laboratory added to NIOSH
1996	Education and Information Division created
2000	Division of Applied Research and Technology formed from the Division of Physical Sciences and Engineering and the Division of Biological and Behavioral Sciences
2001	National Personal Protective Technology Laboratory founded
2006	Respiratory Diseases Research Program founded

Source: NIOSH 2006a.

and Information Division, responsible for health communication and coordinating development of recommendation documents; and the Division of Surveillance, Hazard Evaluations and Field Studies (DSHEFS), responsible for conducting short- and long-term field studies to evaluate health and safety issues including respiratory diseases. DSHEFS is the principal division that handles health hazard evaluations, although the DRDS in Morgantown has also been active in respiratory-disease-related health hazard evaluations.

Other facilities have performed activities related to respiratory disease. One of them is the Health Effects Laboratory Division, which focuses on basic bench laboratory research, including research in basic toxicology and in engineering and exposure assessment. The Pittsburgh Research Laboratory and the Spokane Research Laboratory are mining-focused research groups concerned with research on mining engineering control technology for respiratory hazards. Most recently, the National Personal Protective Technology Laboratory in Bruceton, Pennsylvania, has responsibility for research on respiratory protection and certification of respirators (NIOSH 2006a).

Strategic Planning

Occupational lung diseases have long been a top priority for NIOSH and were the subject of a proposed national strategy in 1986 (NIOSH 1986a). In this strategy, asbestosis, byssinosis (associated with cotton dust), silicosis, and CWP were cited as specific examples for action. The proposed strategy included the following elements: environmental hazard surveillance, medical hazard surveillance, hazard removal, control technologies, regulatory enforcement, worker education and training (includes respirators and other personal protective devices), and worker-oriented programs that include health promotion and smoking prevention (NIOSH 1986a).

In 1996, NIOSH worked with the occupation safety and health community to develop a National Occupational Research Agenda (NORA) (NIOSH 1996a). Seen as a way to better guide and organize research activities in a fiscally constrained environment, NORA identified 21 research priorities grouped into three categories: disease and injury, work environment and workforce, and research tools and approaches (Table 1-2). Under NORA, there were three priorities related to respiratory disease under the disease and injury category: allergic and irritant dermatitis,[1] asthma and COPD, and infectious disease. NIOSH's objectives in creating the NORA approach were to

1. Guide intramural and extramural funding decisions.
2. Encourage and stimulate other government agencies to include NORA priorities in their internal and external research programs.
3. Develop procedures and capacity to track the impact of NORA activities on health and safety outcomes using existing tracking models, if available.
4. Provide for timely updates to the NORA priorities.
5. Periodically review and communicate the overall role and effectiveness of NORA in occupational safety and health.

During development of the NORA agenda, the importance of industrial sector-specific research (for example, construction or agriculture) was consistently raised (NIOSH 1996a). It was decided that the most effective way to integrate this research was through a matrix approach of coordinated research in some or all of the 21 priority areas, as appropriate for each sector. An example of this approach is provided in Table 1-3.

Beginning in 2005, NIOSH introduced the second phase of the NORA process (NORA2), now organizing research priorities focused on eight key industry sectors. The sectors, listed in Table 1-4, are agriculture, forestry, and fishing; construction; health care and social assistance; manufacturing; mining; services; wholesale and retail trade; and transportation, warehousing, and utilities. Sector alignment is intended to provide direct occupational health and safety research and assistance

[1]NIOSH groups allergic and irritant dermatitis in this classification.

TABLE 1-2 NORA Priority Research Areas

Disease and Injury
1. Allergic and irritant dermatitis[a]
2. Asthma and chronic obstructive pulmonary disease[a]
3. Fertility and pregnancy abnormalities
4. Hearing loss
5. Infectious diseases[a]
6. Low back disorders
7. Musculoskeletal disorders of the upper extremities
8. Traumatic injuries
Work Environment and Workforce
9. Emerging technologies
10. Indoor environment
11. Mixed exposures
12. Organization of work
13. Special populations at risk
Research Tools and Approaches
14. Cancer research methods
15. Control technology and personal protective equipment
16. Exposure assessment methods
17. Health services research
18. Intervention effectiveness research
19. Risk assessment methods
20. Social and economic consequences of workplace illness and injury
21. Surveillance research methods

[a]Respiratory disease focus.
Source: NIOSH 1996a.

TABLE 1-3 Example of NORA's Matrix Approach to Coordinating Research

Sector	Allergic and Irritant	Asthma and Chronic Obstructive	Fertility and Pregnancy	Hearing Loss
Agriculture	X[a]	X	X	X
Construction	X		X	X
Service	X	X	X	
Mining	X	X		X
Manufacturing	X	X	X	X

X[a] = priority research area within a sector.
Source: NIOSH 1996a.

TABLE 1-4 NORA2 Structure

NORA Sector Programs
Agriculture, forestry, and fishing
Construction
Health care and social assistance
Manufacturing
Mining
Services
Wholesale and retail trade
Transportation, warehousing, and utilities
NIOSH Cross-Sector Programs
Authoritative recommendations
Cancer, reproductive, and cardiovascular
Communications and information dissemination
Emergency preparedness/response
Global collaborations
Health hazard evaluation
Hearing loss prevention
Immune and dermal
Musculoskeletal disorders
Personal protective technology
Radiation dose reconstruction
Respiratory diseases
Training grants
Traumatic injury
Work organization and stress-related disorders
NIOSH Coordinated Emphasis Areas
Economics
Exposure assessment
Engineering controls
Work life initiative
Occupational health disparities
Small business assistance and outreach
Surveillance

to address specific issues as identified within each sector. Additionally, NORA2 identified 15 cross-sector programs and 7 coordinated emphasis areas, also shown in Table 1-4. The RDRP is one of the cross-sector programs. Each cross-sector program has a steering committee made up of staff from all NIOSH divisions and laboratories as well as the Office of the Director and the Office of Extramural Programs. According to NIOSH (2006a), the steering committee annually reviews and ranks competitive intramural funding requests, reviews all in-house projects relevant to its program, and provides feedback to divisions about each project's relevance and level of priority.

Respiratory Diseases Research Program Structure

As shown in Figure 1-2, the RDRP, as conceptualized, spans nearly all NIOSH units. In 2006, five divisions, three laboratories, and three offices were involved in the RDRP. The mission statement for the RDRP (Weissman 2006) is

> To provide national and international leadership for the prevention of work-related respiratory diseases, using a scientific approach to gather and synthesize information, create knowledge, provide recommendations, and deliver products and services to those who can effect prevention.

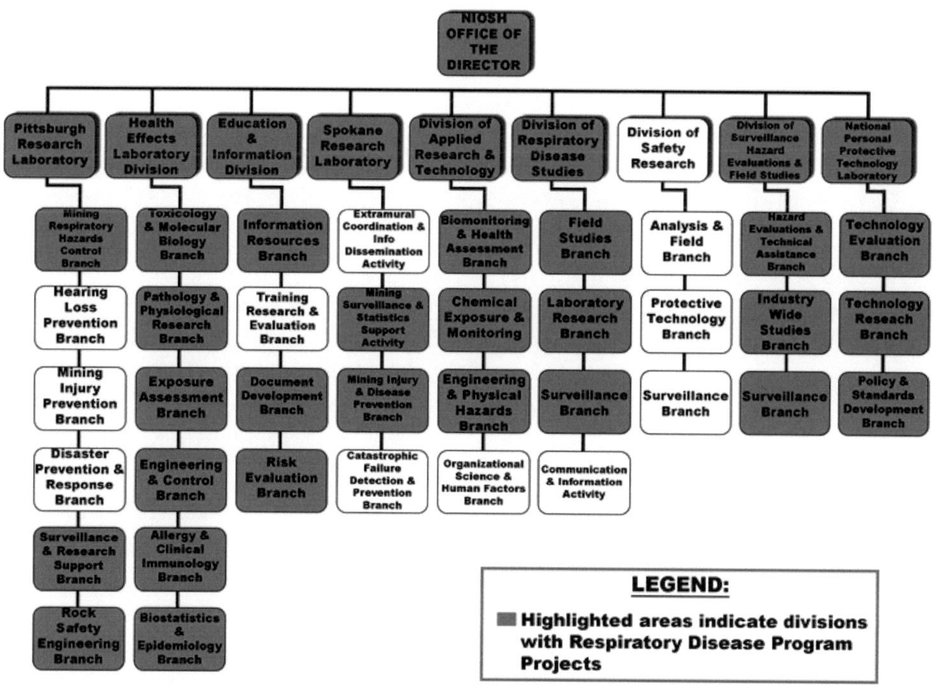

FIGURE 1-2 Respiratory Diseases Research Program projects throughout NIOSH (does not include extramural, communications, research, and technology transfer offices). Source: NIOSH 2006a.

Within this overall mission five specific goals have been stated:

- Prevent and reduce work-related airways diseases.
- Prevent and reduce work-related interstitial lung diseases.
- Prevent and reduce work-related respiratory infectious diseases.
- Prevent and reduce work-related respiratory malignancies.
- Prevent respiratory and other diseases potentially resulting from occupational exposures to nanomaterials.

To pursue these goals, the RDRP activities include multidisciplinary approaches to respiratory diseases through laboratory studies and short- and long-term field studies; surveillance and reporting on disease frequencies; assessment of control technologies and respiratory protection; communication, education, and training; and recommendations on exposure and other practices. Facilities and equipment exist for biomedical research, development of analytical methods, research on exposure assessment, research on engineering and industrial hygiene, respirator research, mining research, and epidemiologic investigations. Ultimately, NIOSH provides the scientific underpinning for development of new or revision of existing standards by OSHA.

NIOSH's Draft Fiscal Year 2008 Program Planning Guide, which describes the process for acquiring competitively awarded intramural research funds at NIOSH, outlines RDRP's strategic and intermediate goals and their performance measures. The document has a high level of specificity, with measurable goals and target time frames. This document became available at the end of the committee's deliberations; thus, the committee has not examined whether they are the correct performance measures or whether they are realistic. The committee notes, however, that because the RDRP is an "organizational unit" borne of the new matrix management approach that spans institutes and divisions, it is unclear whether funding and support are distributed based on whether a project satisfies RDRP's strategic goals or whether RDRP's goals do not matter because funding is distributed based on the goals of the divisions or institutes. Being "at the table" does not provide assurance to the RDRP or to the committee that appropriate weight is given to funding priorities. According to the evidence package, RDRP's steering committee, with representatives from all divisions, "has been empowered to mold activities in respiratory diseases through its abilities to review and rank competitive intramural funding requests and through its access to the NIOSH Office of the Director." (The committee did not attend or review minutes from steering committee meetings.) It is not clear what formal mechanisms exist to ensure that RDRP's goals are supported by funding.

TABLE 1-5 NIOSH RDRP Funding by Fiscal Year, 1996-2005

Year	NIOSH Funding, Millions of Dollars		
	Intramural	Extramural	Total
1996	7.4	2.3	9.7
1997	7.1	3.1	10.2
1998	10.9	4.5	15.4
1999	13.2	7.3	20.5
2000	15.5	7.3	22.8
2001	16.2	8.6	24.8
2002	16.1	10.1	26.2
2003	16.4	10.3	26.7
2004	16.3	8.6	24.9
2005	17.2	11.4	28.6
Totals	136.3	73.5	209.8

Source: Ray Sinclair, NIOSH, unpublished material, April 07, 2007. Table 2-2 shows additional data for extramural funding through 2006, resulting in a total for extramural funding of $88 million for the period 1996-2006.

Program Resources

RDRP funding for the most recent fiscal year available (2005) was close to $29 million. Table 1-5 shows the funding levels for the RDRP intramural and extramural funded research for the years 1996-2005. Intramural funding supports staff salaries and benefits as well as goods and services related to staff research activities. Extramural activities are administered and funded through the NIOSH Office of Extramural Programs in Atlanta (greater detail, including by strategic goal, is provided in Chapter 2). Table 1-6 compares the fraction of the budget spent on intramural versus extramural research for the RDRP with several institutes or centers of the National Institutes of Health (NIH) and shows that the RDRP tends to spend about two-thirds of its budget on intramural projects. This is in contrast to research funding at NIH agencies, including the National Institute of Environmental Health Sciences, that tend to spend most of their research funding on extramural research. Given the relatively small research budget compared with other institutes whose research budgets are typically one to two orders of magnitude higher,[2] the committee considered the limited support allocated to extramural programs unfortunate

[2]For example, the fiscal year 2003 research budgets for NIEHS and NHLBI were approximately $500 million and $2.4 billion, respectively (DSA Pubdata, http://grants.nih.gov/grants/award/trends/icfund9803.html). In contrast, the 2003 research budget for RDRP was approximately $27 million (see Table 1-5).

TABLE 1-6 NIH Versus NIOSH Extramural Funding

	Extramural[a] Percent of Annual Research Budget					
Funding Source	2003	2002	2001	2000	1999	1998
National Center for Complementary and Alternative Medicine	82	78	77	79	—	—
National Cancer Institute	77	75	74	75	75	74
National Center on Minority Health and Health Disparities	55	43	0	—	—	—
National Eye Institute	83	84	87	88	88	87
National Heart, Lung, and Blood Institute	87	87	86	87	86	88
National Institute on Aging	83	83	85	84	84	84
National Institute on Alcohol Abuse and Alcoholism	81	82	84	85	86	85
National Institute of Allergy and Infectious Diseases	79	83	85	87	84	86
National Institute of Arthritis and Musculoskeletal and Skin Diseases	87	87	87	90	90	89
National Institute of Child Health and Human Development	82	84	85	84	83	82
National Institute on Drug Abuse	84	84	86	88	88	88
National Institute on Deafness and Other Communication Disorders	84	85	85	86	87	87
National Institute of Dental and Craniofacial Research	75	74	74	74	75	76
National Institute of Diabetes and Digestive and Kidney Diseases	84	85	85	87	86	85
National Institute of Environmental Health Sciences	71	71	72	79	83	84
National Institute of Mental Health	83	83	84	84	83	82
National Institute of Neurological Disorders and Stroke	85	86	86	86	85	85
National Institute of Nursing Research	92	93	95	93	98	95
National Institute for Occupational Safety and Health Respiratory Diseases Research Program	39	39	35	32	36	29

[a]Extramural includes research grants, research and development, training grants, and fellowships.
Source: Ray Sinclair, NIOSH, unpublished material, April 07, 2007, for NIOSH RDRP; DSA Pubdata (http://grants.nih.gov/grants/award/trends/icfund9803.html) for others.

but probably necessary. Because the committee was not asked to review the extramural program, the committee is not in a position to evaluate the balance between extramural and intramural funding. However, as Table 1-5 shows, the proportion of extramural funding has generally increased with time, from 24% in 1996 to 40% in 2005, the latest year for which complete data are available.

STUDY CHARGE AND EVALUATION COMMITTEE

In September 2004, NIOSH requested that the National Academies review various NIOSH programs with respect to the impact and relevance of their work in reducing workplace injury and illness and to identify future directions that their work might take. The Committee to Review the NIOSH Research Programs was established as an oversight committee and created a framework document, which will be used to evaluate individual programs (Appendix A) (NAS 2005). The RDRP was selected as one of the programs to undergo such a review. The National Research Council (NRC) convened the Committee to Review the NIOSH Respiratory Diseases Research Program in late 2006. The statement of task for the committee is as follows:

> In response to a request from the National Institute for Occupational Safety and Health (NIOSH), the Institute of Medicine (IOM) and the Division of Earth and Life Studies (DELS) of the National Academies (NA) are conducting a series of evaluations of NIOSH research programs. Each evaluation will be conducted by an ad hoc committee, using a methodology and framework developed by the Committee to Review NIOSH Research Programs (framework committee).
>
> Each evaluation committee will review the program's impact, relevance, and future directions. The evaluation committee will evaluate not only what the NIOSH research program is producing, but will also determine whether it is appropriate to credit NIOSH research with changes in workplace practices, hazardous exposures, and/or occupational illnesses and injuries, or whether the changes are the result of other factors unrelated to NIOSH.
>
> The program reviews should focus on evaluating the program's impact and relevance to health and safety issues in the workplace and make recommendations for improvement. In conducting the review, the evaluation committee will address the following elements:
>
> 1. Assessment of the program's contribution through occupational safety and health research to reductions in workplace hazardous exposures, illnesses, or injuries through:
>   a. An assessment of the relevance of the program's activities to the improvement of occupational safety and health, and
>   b. An evaluation of the impact that the program's research has had in reducing work-related hazardous exposures, illnesses, and injuries.

The evaluation committee will rate the performance of the program for its relevance and impact using a scale of 1 to 5. Impact may be assessed directly (e.g., reductions in illnesses or injuries) or, as necessary, using intermediate outcomes to estimate impact. Qualitative narrative evaluations should be included to explain the numerical ratings.

2. Assessment of the program's effectiveness in targeting new research areas and identifying emerging issues in occupational safety and health most relevant to future improvements in workplace protection. The committee will provide a qualitative narrative assessment of the program's efforts and suggestions about emerging issues that the program should be prepared to address.

The study committee was selected to include members with expertise in epidemiology, exposure assessment, industrial hygiene, inhalation toxicology, occupational medicine, and pulmonology. The committee reviewed the RDRP emphasizing, although not exclusively limited to, the time period since 1996, as this time frame encompasses both the original NORA and the recently completed NORA2. However, the committee is cognizant that improvements in workers' health with regard to respiratory diseases that occurred during this period may be a result of NIOSH research activities that were completed much earlier. The committee met three times in the period October 2006 through March 2007. The first two meetings were data-gathering sessions that included presentations by NIOSH staff and other invited speakers in open session. At the end of each open session stakeholders and the general public had an opportunity to comment. In addition, several committee members and a member of the NRC staff attended a site visit at the NIOSH Morgantown, Cincinnati, and Pittsburgh facilities.

The committee's review of the NIOSH RDRP was based in part on materials provided by NIOSH at the onset and on specific materials the committee requested during the review (see Appendix C for a list of these materials). The committee wanted to hear from a broad range of stakeholders and created an online questionnaire for them to provide comments (see Appendix B). Presentations and discussions during the open session as well as the online questionnaire helped in shaping additional questions and giving background to areas under review.

EVALUATION APPROACH

Framework Document and Logic Model

The committee evaluated the relevance of RDRP research to improvements in occupational safety and health and the impact that NIOSH research has had in reducing occupationally related morbidity. In these efforts, the committee used the framework document (Appendix A) developed by the Committee for the Review of NIOSH Research Programs to provide structure and guidance to the individual evaluating committees (NAS 2005). Figure 1-3 outlines the general approach the framework document suggests using in evaluating the relevance and impacts of the RDRP.

Relevance was evaluated in terms of the degree of research priority and connection to improvements in workplace protection. Factors taken into account include the frequency and severity of health outcomes and the number of people at risk, the structure of the program, and the degree of consideration of stakeholders' input. The impact of the program's research is evaluated in terms of its contributions to worker health and safety. The framework document recommends that the evaluation committee look at each strategic goal and present the committee's assessments with regard to the relevance and impacts of the research completed or in progress for each separate goal. At the end of the review of the five strategic goals, the committee then develops a quantitative score for the relevance and impact of

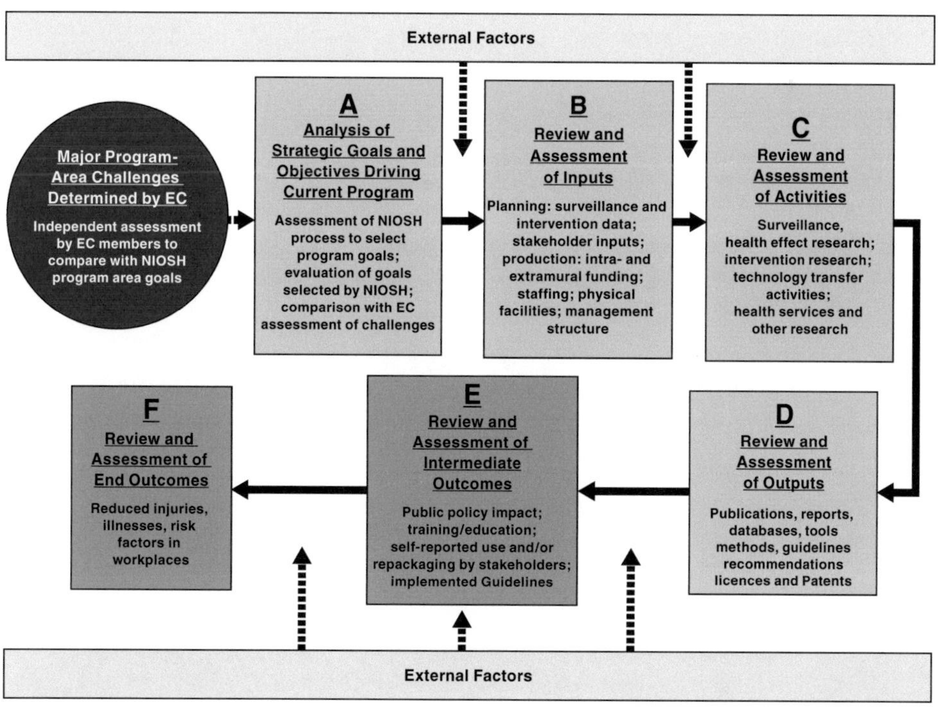

FIGURE 1-3 Flowchart for evaluation of the NIOSH research program.

the program as a whole and provides its quantitative and qualitative evaluation of the relevance and impacts of the RDRP. The evaluation is to conclude with detailed qualitative assessments as well as the assignment of scores between 1 and 5 for the relevance and impact of the RDRP research and other activities. Box 1-1 and Box 1-2 provide the scoring criteria for rating the programs.

The study charge also directs the committee to review the progress the RDRP has made in identifying new research and provides the committee the opportunity to identify emerging research areas relevant to the program's mission. According to the framework document, the committee's identification of emerging research areas was done using members' expert judgment rather than a formal research needs identification effort.

The guidance in the framework document reflects the terminology and organization of a logic model adopted by NIOSH to characterize the steps in its work. The logic model used by the RDRP is shown in Figure 1-4, and examples of the terms used within the logic model are provided in Box 1-3. The committee reached consensus on its assessment of the individual activities of the RDRP and the program as a whole through deliberations at meetings and discussion on its written materials. This included a portion of one meeting spent devoted solely to develop-

BOX 1-1
Criteria for Rating Relevance

5 = Research is in highest-priority subject areas and highly relevant to improvements in workplace protection; research results in, and NIOSH is engaged in, transfer activities at a significant level (highest rating).

4 = Research is in highest-priority subject areas and adequately connected to improvements in workplace protection; research results in, and NIOSH is engaged in, transfer activities.

3 = Research focuses on lesser priorities and is loosely or only indirectly connected to workplace protection; NIOSH is not significantly involved in transfer activities.

2 = Research program is not well integrated or well focused on priorities, is not clearly connected to workplace protection, and is inadequately connected to transfer activities.

1 = Research in the research program is an ad hoc collection of projects, is not integrated into a program, and is not likely to improve workplace safety or health.

BOX 1-2
Criteria for Rating Impact

5 = Research program has made a major contribution to worker health and safety on the basis of end outcomes or well-accepted intermediate outcomes.

4 = Research program has made a moderate contribution on the basis of end outcomes or well-accepted intermediate outcomes; or research program generated important new knowledge and is engaged in transfer activities, but well-accepted intermediate outcomes or end outcomes have not been documented.

3 = Research program activities or outputs are going on and are likely to improve worker health and safety (with explanation of why not rated higher). Transfer activity is planned.

2 = Research program activities or outputs are going on and may result in new knowledge or technology, but only limited application is expected. NIOSH is not significantly involved in transfer activities.

1 = Research activities and outputs are NOT likely to have any application.

NA = Impact cannot be assessed; program is not mature enough.

ing scores for relevance and impact. Assessments of "goals," "inputs," "activities," and "outputs" were used to evaluate the relevance of the program's research. End and intermediate outcomes were the principal focus for evaluation of the impact of the program's research. The committee's scores for relevance and impact are presented and discussed in Chapter 2 of this report.

Information Sources

The material for this review comes from: (1) the written documentation, known as the evidence package, provided by NIOSH to the evaluation committee before the first meeting of the evaluation committee (NIOSH 2006a); (2) presentations by NIOSH and stakeholders at the committee meetings; (3) responses to letters sent to NIOSH by the evaluation committee dated November 17, 2006, and December 6, 2006; (4) review of appendix material submitted with the original written documentation and a presentation at the October 26, 2006, meeting of the evaluation committee. The committee also drew on individual members' knowl-

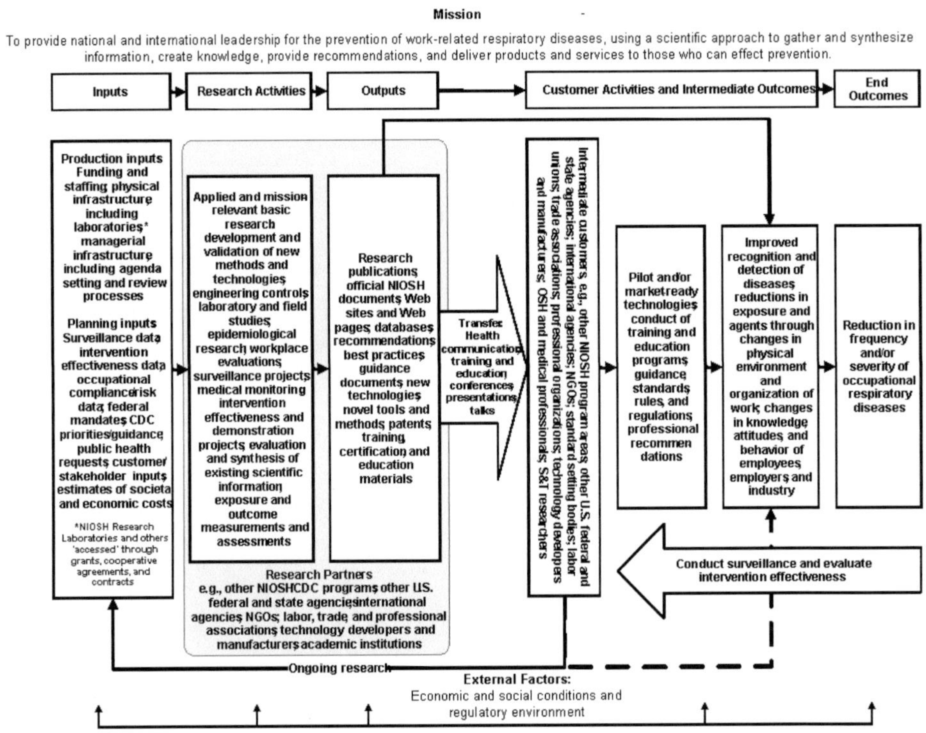

FIGURE 1-4 Logic model for the RDRP. Abbreviations: CDC, Centers for Disease Control and Prevention; NGOs, nongovernmental organizations; OSH, Occupational Safety and Health. Source: NIOSH 2006a.

edge of NIOSH's respiratory disease research and on standard literature searches (e.g., PubMed and other databases) where appropriate.

ORGANIZATION OF THE REPORT

The remainder of the report presents the findings from the committee's evaluation. Chapter 2 presents the committee's review of the NIOSH RDRP and the ratings for the program's relevance and impact in reducing workplace injury and illness. In Chapter 3, the committee reviews the RDRP's mechanisms for identifying emerging issues in occupational respiratory disease and identifies issues that may warrant future attention. In Chapter 4, the committee identifies opportunities to strengthen the NIOSH RDRP and increase the relevance and impact of the program's efforts.

BOX 1-3
Logic Model Terms and Examples

Planning Inputs: Stakeholder input, surveillance and intervention data, and risk assessments.

Production Inputs: Intramural and extramural funding, staffing, management structure, and physical facilities.

Activities: Efforts and work of the program, staff, grantees, and contractors.

Outputs: A direct product of a NIOSH research program that is logically related to the achievement of desirable and intended outcomes.

Intermediate Outcomes: Related to the program's association with behaviors and changes at individual, group, and organizational levels in the workplace. An assessment of the worth of NIOSH research and its products by outside stakeholders.

End Outcomes: Improvements in safety and health in the workplace. Defined by measures of health and safety and of impact on processes and programs.

External Factors: Actions of forces beyond NIOSH's control with important bearing on incorporation in the workplace of NIOSH's outputs to enhance health and safety.

2

Evaluation of the Respiratory Diseases Research Program

The committee was charged with reviewing the Respiratory Diseases Research Program (RDRP) of the National Institute for Occupational Safety and Health (NIOSH) to evaluate the relevance and impact of its work to reduce workplace illnesses and injuries. The committee's review focused primarily on 1996 to the present. This time period encompasses not only the current RDRP as it was established under the second National Occupation Research Agenda (NORA2) in 2005 but also occupational respiratory disease research as it was configured under the original NORA in 1996. The committee also notes selected activities related to respiratory disease research that occurred before 1996. As stated previously, the committee's use of the term RDRP also indicates NIOSH activities and programs related to occupational respiratory diseases that predate formal creation of the RDRP under NORA2.

The committee evaluation followed the framework document, presented in Appendix A, developed by the Committee to Review NIOSH Research Programs and referred to as the framework committee. The framework document directs individual evaluation committees to evaluate program relevance in terms of the degree to which the research is connected to improvements in workplace protection. It identifies factors to be considered, including the frequency and severity of adverse health outcomes, the number of people at risk, the structure of the program, and the degree of consideration of stakeholder inputs. The framework document directs the evaluation committee to evaluate research impact in terms of the program's contributions to improvement in worker health and safety to the extent that it can be known or surmised in terms of quantifiable outcomes. This chapter presents the results of the committee's review.

RDRP GOALS, SUBGOALS, AND RESOURCES

Since 2006, the RDRP has pursued the five strategic goals discussed in Chapter 1 with activities of different breadth. These goals are further subdivided into the subgoals shown in Table 2-1. NIOSH used these five strategic goals and their related subgoals to organize the primary evidence package and presentations provided to the committee. In turn, the committee used the five goals to organize its evaluation of the RDRP. While recognizing that the program did not directly use these goals during most of the period covered by the committee's assessment, the goals are nonetheless consistent with priorities adopted by NIOSH throughout its history. These goals are also relevant for work related to the NORA priority areas of control technology and personal protective equipment, exposure assessment methods, and intervention effectiveness research. The following sections present the committee's findings addressing both the overall program and matters concerning individual goals and their related subgoals.

Funding for the program goals has varied over the past 10 years. Table 2-2 provides the budget for the RDRP classified by program goal. The largest fraction of the RDRP budget goes to airway diseases, followed closely by research budgeted for interstitial lung diseases. Table 2-2 also shows that malignancies and nanotechnologies are the smallest components, although spending on nanotechnology research is rapidly increasing. Figure 2-1 shows a trend toward research in airway and interstitial disease and away from the study of occupational respiratory malignancies.

EXTERNAL FACTORS AFFECTING THE RDRP

The RDRP operates in an environment shaped by many factors that the program cannot control. Some factors are so fundamental to the nature of the program that the committee found it essential to keep them in mind for all aspects of its review. For example, NIOSH is primarily a research entity. Thus, although NIOSH can issue recommendations, it cannot mandate that such recommendations be implemented in the workplace. Regulatory implementation of NIOSH's recommendations is left to the Occupational Safety and Health Administration (OSHA) or the Mine Safety and Health Administration (MSHA).

Another important consideration is that the RDRP comprises a collection of activities that take place within the 11 organizational units of NIOSH shown in Table 2-3. The program is based on a matrix approach and does not reside within a

TABLE 2-1 Strategic Goals and Subgoals of the NIOSH RDRP, as of February 2007

Strategic Goal 1: Prevent and reduce work-related airway diseases
- Prevent and reduce asthma and allergy due to natural rubber latex among health care workers
- Prevent and reduce WRA in the isocyanate production industry[a]
- Prevent and reduce WRA related to nonindustrial indoor environmental quality
- Improve detection of WRA and relevant exposures
- Establish the work-relatedness of COPD
- Develop tools and identify at-risk workers in industries and occupations to assess the extent, severity, and burden of work-related COPD
- Develop, test, and disseminate recommendations for preventing COPD in the workplace.
- Prevent and reduce flavoring related bronchiolitis obliterans

Strategic Goal 2: Prevent and reduce work-related interstitial lung diseases
- Prevent and reduce respiratory diseases induced by coal mine dust
- Prevent and reduce silica-induced respiratory diseases
- Prevent and reduce fiber-induced diseases
- Prevent and reduce chronic beryllium disease

Strategic Goal 3: Prevent and reduce work-related infectious respiratory diseases
- Maintain reductions in the incidence of occupational tuberculosis in high-risk work settings
- Protect workers from bioterrorism agents
- Protect workers from occupational acquisition of emerging diseases (including severe acute respiratory syndrome, avian and pandemic flu)
- Protect workers from occupational exposures that make them susceptible to respiratory infections
- Prevent outbreaks of occupational histoplasmosis by maintaining worker and employer awareness

Strategic Goal 4: Prevent and reduce work-related respiratory malignancies
- Determine occupation etiologies of lung cancer
- Reduce metal-induced lung cancer (hexavalent chromium)
- Prevent and reduce silica-induced lung cancer
- Prevent and reduce lung cancer induced by diesel engine exhaust
- Produce lung cancer diagnostic tools

Strategic Goal 5: Prevent respiratory and other diseases potentially resulting from occupational exposures to nanomaterials
- Determine the relative toxicity of nanomaterials
- Conduct exposure assessment and engineering control evaluations in 10 nanomaterial production or use facilities by 2008
- Produce dose-response data for carbon nanotubes sufficient to conduct a quantitative risk assessment by 2008

Abbreviation: WRA, work-related asthma.

[a]Because workers are exposed to isocyanates in a variety of industries besides isocyanate production, the goal should probably be worded "Preventing and reducing WRA associated with isocyanate exposure." However, to be consistent with the NIOSH terminology, NIOSH's original wording is used in this report.

TABLE 2-2 NIOSH Respiratory Diseases Research Program Budget and Staffing by Research Goals (in Millions of Dollars)

Total	FY1996	FY1997	FY1998	FY1999	FY2000	FY2001	FY2002	FY2003	FY2004	FY2005	FY2006
Intramural	7.35	7.09	10.90	13.20	15.50	16.20	16.06	16.39	16.31	17.17	
Extramural	2.31	3.08	4.51	7.31	7.29	8.61	10.08	10.30	8.64	11.40	14.50
Strategic Goal 1: Prevent and reduce work-related airway diseases											
Intramural	1.66	1.56	2.07	4.35	6.38	5.90	5.64	5.97	7.31	5.80	
Extramural	0.62	1.51	2.37	3.13	3.72	3.42	3.10	2.83	2.52	2.74	2.94
Strategic Goal 2: Prevent and reduce work-related interstitial lung diseases											
Intramural	1.73	3.56	4.60	5.61	6.82	6.14	6.25	6.60	5.61	6.64	
Extramural	0.08	0.08	0.35	0.70	0.65	1.48	0.92	1.09	0.38	0.68	0.38
Strategic Goal 3: Prevent and reduce work-related infectious respiratory diseases											
Intramural	2.89	1.74	3.57	1.89	1.40	3.10	3.63	3.05	2.45	2.70	
Extramural	0.30	0.50	0.65	0.65	0.29	0.26	0.25	0.00	0.08	0.08	0.75
Strategic Goal 4: Prevent and reduce work-related respiratory malignancies											
Intramural	1.08	0.23	0.66	1.34	0.90	1.06	0.54	0.76	0.74	0.76	
Extramural	.043	0.19	0.09	1.68	1.68	1.19	1.25	1.91	1.78	1.40	1.35
Strategic Goal 5: Prevent respiratory and other diseases potentially resulting from occupational exposures to nanomaterials											
Intramural	0	0	0	0	0	0	0	0	0.67	1.27	3.31
Extramural					0.10	0.38	0.38	0.29	0.10	0.66	0.82

Abbreviations: FY, fiscal year; FTE, full-time equivalent.
Source: R. Sinclair, NIOSH, unpublished material, April 7, 2007.

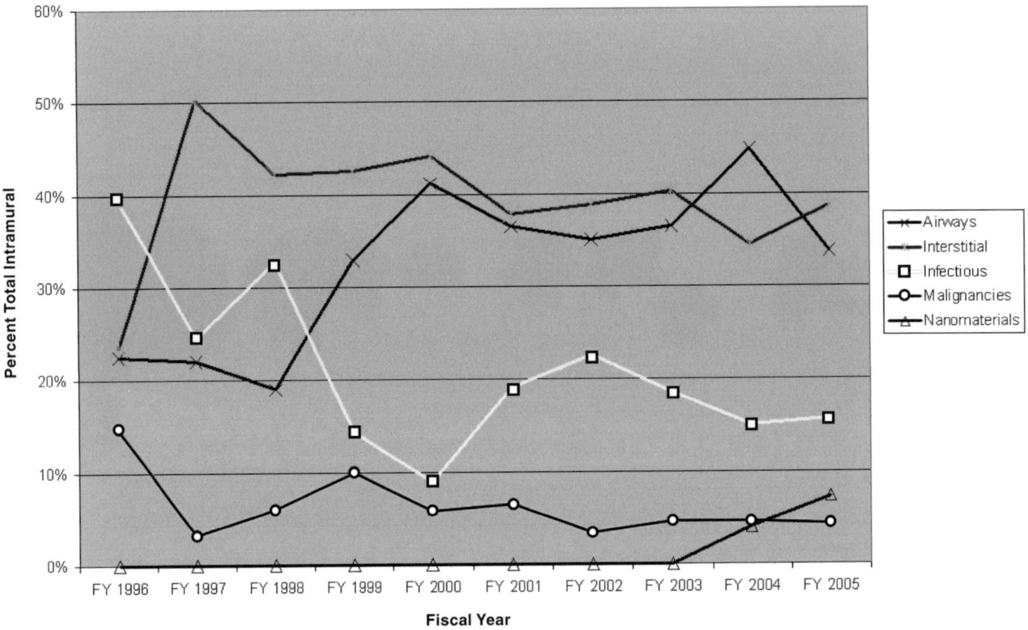

FIGURE 2-1 Proportional allocation of RDRP intramural funds by goal by fiscal year (FY). Source: R. Sinclair, NIOSH, unpublished material, April 7, 2007.

specific division or laboratory on the NIOSH organizational chart (see Chapter 1). Although there is a Division of Respiratory Disease Studies, it is only one part of the RDRP, albeit the leading one. As with all 15 cross-sector programs under NORA2, a program manager and coordinator(s) are expected to monitor and guide the overall program effort. However, as described in Chapter 1, the organizational structure of NIOSH means that the program manager of the RDRP does not control the budget or the entire program portfolio. The program's funding level is the sum of the financial resources that individual NIOSH organizational units apply to work on activities to monitor and prevent respiratory disease. The activities of the intramural program are distributed across units located in Morgantown, West Virginia; Pittsburgh, Pennsylvania; Cincinnati, Ohio; and Spokane, Washington. The selection and management of extramural projects are based in Atlanta, Georgia.

The portfolio, staffing, and funding levels for the RDRP also are shaped by congressional direction as to the amount of the NIOSH budget to be applied to priority areas (e.g., mining safety and health). Although the NIOSH mining program no longer carries a separate line item in the federal budget, Congress has directed

TABLE 2-3 NIOSH Divisions, Laboratories, and Offices Involved in the RDRP

- Division of Respiratory Disease Studies
- Health Effects Laboratory Division
- Division of Surveillance Hazard Evaluation and Field Studies
- Division of Applied Research and Technology
- Education and Information Division
- Pittsburgh Research Laboratory
- Spokane Research Laboratory
- National Personal Protective Technology Laboratory
- Office of Extramural Programs
- Office of Health Communications
- Office of Research and Technology Transfer

NIOSH to maintain its current level of research effort in this area. In practical terms, this means that, although the RDRP receives a percentage of its budget from the mining sector program, it does not have discretion to redirect these funds to any of the program's other activities, which may need funding, unless there is a nexus with mine safety and health.

SURVEILLANCE, HEALTH HAZARD EVALUATION AND TECHNICAL ASSISTANCE PROGRAM, AND OTHER INPUT ACTIVITIES

There are several NIOSH programs that support all NIOSH activities, including those of the RDRP. These programs include surveillance activities, the health hazard evaluation and technical assistance (HHE/TA) program, emergency response and disaster preparedness tasks, and the respirator program. These elements serve as critical inputs into and support for the RDRP. Because they serve many different NIOSH activities, they are not evaluated under the program goals of the RDRP. However, it is important to note how these activities help serve the RDRP.

Surveillance

WoRLD Reports

The information gathered through surveillance helps to identify and track occupational health and exposure. This information also can provide research directions and allow for assessment of trends and impacts. One prominent surveillance activity is the summary of occupation respiratory disease and associated exposure data in the series of Work-Related Lung Disease (WoRLD) Surveillance Reports and the web-based eWoRLD Surveillance System. The WoRLD Surveillance

Report provides information on the frequency and trends in respiratory diseases, the industry and geographic distribution, and demographic information on those affected. Six of these reports have been released since 1991. As noted by NIOSH (2006a), the continuing challenge of such an undertaking not only is to maintain and update existing content but also to present the content in a meaningful way and to add new sources. This activity has also helped spur work in developing a standardized approach to mortality data on occupational lung disease. It has also helped give rise to NIOSH's *Worker Health Chartbook* (NIOSH 2004).

SENSOR Programs

The SENSOR program attempts to integrate occupation health surveillance into public health activities at the state level. Using a consensus case definition of occupational asthma, SENSOR began as a sentinel event system based solely on voluntary physician reports. It has evolved into efforts at the state level to enhance the population aspects of work-related asthma (WRA) surveillance including examination of hospital discharge data and use of capture-recapture methodology. The state surveillance programs were intended to be targeted programs. At its inception in 10 states in 1987, 6 states identified WRA and 4 identified silicosis for surveillance. Currently, only 4 states continue WRA surveillance (Massachusetts, Washington, Michigan, and California) and 2 continue silicosis surveillance (Michigan and New Jersey).

The need for appropriate surveillance data for WRA is critical. Without such data, the appropriate targeting of limited resources for exposure and medical monitoring and interventions to prevent and control disease is difficult, if not impossible. The SENSOR program now has a presence in only four states. Expanding the program to include additional geographic areas, and thus the broader occupational mix, would provide a larger basis to track trends and to highlight geographic variation. It is difficult to confirm from external material the NIOSH conclusion that the RDRP "indirectly supports 33 states, one city, and one territory to conduct surveillance for WRA." While the National Center for Environmental Health supported state asthma plans and encouraged recipient state health departments to develop methods for WRA surveillance, a nationwide standard was not implemented and data formats and quality vary.

Health Hazard Evaluation and Technical Assistance Program

Under its "right of entry" authority, NIOSH investigates suspected hazardous exposures in the workplace. Such investigations, known as health hazard evaluations, take place under the HHE/TA program. These investigations are initiated by requests from workers, from their representatives, or from an employer. Because of the importance and unique character of the HHE/TA program, it will be the subject of a separate National Research Council evaluation. This section focuses on the role of the HHE/TA program insofar as it interfaces with the goals of the RDRP.

As noted later in this chapter, particularly with regard to activities related to airways and infectious diseases, the RDRP appears to have effectively used the HHE/TA program to inform and extend the research program on workplace environmental quality. For example, the HHE of the Missouri microwave popcorn plant, which investigated flavoring-induced bronchiolitis obliterans, gave rise to a coordinated research program on flavoring-related obstructive airway disease. The response to that specific HHE request can be considered a model of how the surveillance and initial data-gathering activities related to an HHE can be used as a key production input for setting priorities and planning research objectives.

A more systematic approach to the review of all investigations in the HHE/TA program related to respiratory diseases might prove useful. For example, when the outbreak of flavoring-related bronchiolitis obliterans was followed up it was learned that an HHE a decade before had examined a similar work risk. No systematic assessment of HHE findings appears to be in place to identify other new occupational respiratory disease risks worthy of investigation. Similarly, the committee did not learn of any general review of reports in the HHE/TA program for WRA due to low-molecular-weight (LMW) or high-molecular-weight (HMW) sensitizers and irritants that induce new-onset asthma or for the discovery of agents that may cause work-related interstitial or airway disease, although NIOSH did undertake such a summary report with regard to isocyanates (Donovan Reh 2004).

Dissemination of particularly relevant outbreak investigations pertinent to WRA or fixed-airway obstruction through mechanisms such as the *Morbidity and Mortality Weekly Report* (MMWR) or brief case reports in the peer-reviewed literature also could be carried out more effectively (e.g., glutaraldehyde in heart-valve manufacturing workers, chlorine/sulfur dioxide/ozone in paper and pulp mill workers). A successful example of such dissemination by NIOSH was a report of a cluster of cases of new-onset asthma associated with exposure to 3-amino-5-mercapto-1,2,4-triazole identified through SENSOR (Hnizdo et al. 2004a) surveillance.

Emergency Response

As described in the evidence package, NIOSH participates in agency-wide responses to emergencies and catastrophes. These events include the World Trade Center disaster and anthrax attacks, responses to hurricanes Katrina and Rita, and responses to outbreaks of severe acute respiratory syndrome (SARS) and avian influenza in 2005. Activities related to emergency responses include immediate activities related to specific incidents and follow-up activities to further assess the impacts of these incidents and to help improve preparedness. Such activities are incorporated into RDRP program goals to varying degrees. For example, activities related to the goal of preventing and reducing work-related infectious respiratory diseases have been informed by the anthrax attacks of 2001 and outbreaks of SARS and avian influenza. However, no research activities related to emergency response and either WRA or chronic obstructive pulmonary disease (COPD) were reported in the airways disease section of the evidence package. Interestingly, the evidence package did not cite the ongoing medical monitoring of the World Trade Center disaster rescue and recovery workers as an example. This may reflect the somewhat compartmentalized organizational structure of NIOSH. It is unclear whether the RDRP plans to incorporate such research questions into future efforts, but such events could lead to coordinated activities between intramurally and extramurally funded NIOSH research programs.

Respiratory Program

Another critical element of NIOSH activities relevant to respiratory diseases is work related to respirators, including respirator policy. The evidence package provided by NIOSH (NIOSH 2006b) provides an overview of these activities, including the role NIOSH plays in having the lead responsibility for directing and carrying out the NIOSH respirator certification program and related laboratory, field, quality, and research activities. However, the integration of this activity with the program goals of the RDRP was unclear. For example, there was no discussion of airway-disease-specific respirator policy in the evidence package. However, NIOSH respirator development, testing, and certification efforts continue to play a crucial role in preventing both WRA and work-related COPD. The agency could better highlight the need for the respirator program by using the recent NIOSH peer-reviewed report supporting the use of respirators in emergency response situations that involve exposures to irritant dust (Feldman et al. 2004).

INTRODUCTION TO RELEVANCE AND IMPACTS ASSESMENT

The following sections review the five NIOSH RDRP strategic goals and present the committee's assessments with regard to the relevance and impacts of the research completed or in progress. The committee's evaluation of NIOSH's targeting of new research is discussed in Chapter 3. At the end of the review of the five strategic goals, the committee provides a quantitative and qualitative evaluation of the relevance and impacts of the RDRP. In its evaluation of the relevance of the work of the NIOSH RDRP, the committee has assessed the degree to which the program has led and carried out research in aspects of occupationally related respiratory disease most relevant to improvements in workplace protection. As discussed in Chapter 1, the committee evaluated the RDRP by using the evidence presented by NIOSH and other stakeholders, a review of the literature, and committee members' knowledge and experience in respiratory diseases. The committee developed a consensus through deliberations at meetings and discussions of written materials, including a lengthy scoring discussion.

STRATEGIC GOAL 1: PREVENT AND REDUCE WORK-RELATED AIRWAY DISEASES

Introduction

Obstructive airway diseases as a group comprise the most prevalent type of occupationally related chronic respiratory disease. Moreover, the contribution of occupational factors to the incidence of obstructive airway diseases is well documented. Thus, the overall goal of the RDRP airway diseases component to prevent and reduce work-related airway diseases is extremely important. The supporting materials the RDRP submitted to the committee underscore the importance of this goal to NIOSH. Our assessment focuses on the ways program development and associated resource allocation can optimize the efforts being expended to achieve this goal given its relative importance to the overall NIOSH agenda.

The materials devoted to airway diseases submitted by NIOSH were organized into two major disease categories: WRA and fixed obstructive airway diseases (COPD and bronchiolitis obliterans). Four specific objectives were listed by NIOSH for each of these two categories. To facilitate review, the evaluation of the relevance and impact of the RDRP airway diseases component activities follows the same organizational scheme and addresses separately the subgoals for WRA and fixed airway obstruction as formulated by NIOSH. Nonetheless, it is recognized that no rigid boundary demarcates these two broad categories of obstructive airway disease and substantial overlap can occur.

Work-Related Asthma

NIOSH formulated the following four subgoals for WRA:

- Prevent and reduce asthma and allergy due to natural rubber latex among health care workers.
- Prevent and reduce WRA in the isocyanate production industry.

- Prevent and reduce WRA related to nonindustrial indoor environmental quality.
- Improve detection of WRA and relevant exposures.

Ample rationale exists for including such specific objectives in an overall WRA program. The prevalence of sensitization to natural rubber latex among health care workers increased dramatically between 1987 and 1996 because of a huge increase in the use of latex gloves as part of the universal precautions for preventing exposure to blood-borne pathogens. Diisocyanates are one of the most common specific chemical causes of WRA and are used in a variety of settings and formulations. Indoor environmental quality complaints in general, and complaints specifically related to respiratory health (which include exacerbation of preexisting asthma), have led to multiple NIOSH HHEs. In response, NIOSH has developed considerable expertise related to WRA exacerbation in the context of indoor air, largely in the managerial and service sector, which is a dominant employment sector in postindustrial economies. The fourth, more general objective recognizes that improved surveillance for WRA is critical for identifying emergent problems, setting priorities, and tracking the effectiveness of interventions.

The objectives as formulated have certain limitations. The objective related to latex asthma and allergy largely has been achieved. NIOSH deserves much credit for the documented reduction in WRA due to natural rubber latex, an end outcome among health care workers in the United States. The case for continuing this narrowly cast objective as a priority could have been more clearly elucidated. Latex is still used widely in applications far beyond the health care industry, such as day care and other service industries. Programmatic work on latex, a prototypical HMW antigen, naturally extends to a variety of other important causes of WRA, such as enzyme and nonenzyme plant and animal proteins. This work also is relevant to work-related rhinitis, an airway disease apparently overlooked in the description of NIOSH goals.

Similarly, the importance of diisocyanate-related asthma, a key prototype for asthma induced by LMW chemicals, goes well beyond the manufacture of this chemical (the focus of the NIOSH goal). In particular, diisocyanates are relevant to end-user sectors, especially in the construction industry. The mechanisms of immune response to these chemicals appear to differ in key ways from classic HMW sensitization, such as sensitization to latex, so the diisocyanate-specific goal may also be generalized to a far wider array of workplace LMW chemical hazards. For example, the challenges of elucidating mechanisms of disease in isocyanate-induced asthma may apply to other LMW chemicals. The NIOSH supporting materials only minimally addressed this subject.

Work-aggravated asthma only recently has been recognized as a major issue in occupational airway disease, in large part through NIOSH's leadership role. Examination of indoor air quality in offices and other public buildings, highlighted in the WRA goal related to asthma exacerbations, may well be expanded to other settings, such as custodial work.

Finally, the admirable, but broadly stated generic goal of improved detection, suffers when juxtaposed with the other highly focused objectives. NIOSH's crafting of this goal could have benefited from concentrating on selected key examples. For example, the targeted surveillance of irritant-induced asthma is overlooked although highly appropriate to this objective. Whatever shortcomings exist in the definitions of this key goal, NIOSH has been suitably active in this area. For example, a NIOSH investigator is one of the coauthors of the chapter on this subject in a major text in the field and has been involved in important investigations of the subject (Gautrin et al. 2006).

Planning and Production Inputs

NIOSH-funded surveillance efforts through the SENSOR program assisted in determining the extent of the latex-induced occupational asthma problem and, importantly, identified its prominence relative to other HMW antigens for which latex can serve as a prototype. The established surveillance system can continue to provide extended tracking information for latex-caused WRA, allowing for assessment of intervention efforts. Although limited to four states and with an uncertain future, SENSOR for WRA has also facilitated the identification of emerging problems associated with other HMW antigens. Additionally, the RDRP developed an approach to the latex problem that included collaboration with multiple federal agencies, health care industry associations, and professional organizations as well as the preparation of a widely disseminated NIOSH Alert (see below).

Diisocyanates are widely used in U.S. industry and multiple studies have shown them to be one of the most commonly identified causes of LMW sensitizer-induced occupational asthma. There is no reliable biological marker of sensitization to diisocyanates and the OSHA permissible exposure limit appears to be inadequate. The RDRP recognizes both the relative importance of diisocyanates as a cause of asthma and the relative difficulty of detecting diisocyanate-induced asthma or sensitization at the preclinical stage. The RDRP approach to the diisocyanate problem includes supporting both intramural and extramural research, developing new exposure monitoring methods, partnering with industry to develop best practices for exposure control and medical monitoring, and using surveillance through the SENSOR program.

As highlighted above, available data suggest that improving the environmental quality of nonindustrial indoor workplaces can have major health and economic benefits. Review of the HHE program showed that there has been a tremendous increase in requests for HHEs related to indoor environmental quality concerns over the past three decades, with building-related asthma accounting for approximately 25% of such requests. An Institute of Medicine report (IOM 2004) documented that damp indoor spaces are associated with health risks and public health efforts should be directed toward preventing and remediating such environments. Data from the SENSOR program also identified indoor air pollutants as a category of agents frequently associated with WRA. The RDRP developed an approach to this problem that resulted in a major investment of intramural funds through the Asthma Research Program Project. The RDRP also developed collaborations in Connecticut that involved state agencies and the University of Connecticut in conducting a multiyear study of a large office building. In Maine, state agencies and the American Lung Association Maine affiliate work with the RDRP to study the relationship between school environmental quality and asthma. Other collaborations were formed with universities and the U.S. Environmental Protection Agency (EPA) to develop methods for assessing exposures to fungi.

The asthma and COPD NORA1 team recognized the need for an improved questionnaire to ascertain probable WRA in epidemiologic studies. The RDRP also recognized the need for a portable spirometer, a personal device to record and store data electronically, that could be used to improve the detection of WRA. The RDRP developed collaborative approaches to these needs that involved working with the American Thoracic Society (ATS) and the National Center for Health Statistics (NCHS) on questionnaire development and the manufacturer of a popular portable spirometer to improve the device for use in detecting WRA.

The planning and consideration of inputs for all four subgoals appear to have been reasonable. The asthma and COPD team during NORA1 was aware of the problems of latex- and diisocyanate-induced occupational asthma and of the larger problem of inadequate surveillance for WRA. The team devoted its efforts toward addressing these and other asthma-related research and prevention issues. It is not clear how well the new industrial-sector-based approach of NORA2 will be able to address disease-specific issues. The indoor environment team during NORA1 provided a forum for addressing problems of indoor air quality common to office and other buildings in virtually all areas of the economy. Again, it is not clear how the NORA2 approach will address disease-related problems that are common across industrial sectors.

Activities, Outputs, and Outcomes

The primary output of the natural rubber latex activity was the publication of a NIOSH Alert, "Preventing Allergic Reactions to Natural Rubber Latex in the Workplace" (NIOSH 1997a). More than 360,000 hard copies have been disseminated since it was released in 1997, with easy public access to a downloadable web version. The NIOSH Alert also led to published commentaries in at least 10 health publications accessible on Pubmed (National Library of Medicine) in 1997. In addition, the RDRP conducted a research project on the efficacy of the NIOSH Alert for changing hospital policy that documented its effectiveness and led to two peer-reviewed publications (Maxfield et al. 1999, 2000). The latex reports in the SENSOR program have been analyzed systematically: a peer-reviewed publication based on SENSOR data focuses on WRA among health care workers and highlights latex among other exposures (Pechter et al. 2005). There have been other relevant information transfer efforts as well. An RDRP collaboration with the Veterans Administration health care system demonstrated that switching from powdered to nonpowdered latex gloves dramatically reduced the rate of latex sensitization (Zeiss et al. 2003). An RDRP collaboration with the Food and Drug Administration (FDA) resulted in a peer-reviewed publication that analyzed FDA MedWatch data relevant to adverse events from medical gloves, primarily focusing on latex allergy (Dillard et al. 2002).

Although NIOSH has never published a report specific to latex allergy in the MMWR, an outbreak of occupational asthma related to another HMW trigger (egg protein) was published (MMWR 1987). Similarly, although NIOSH has not committed substantive resources to extramural or intramural research into the biological mechanisms or outcomes of latex-induced occupational asthma, it has supported work related to allergic sensitization to egg protein and asthma associated with crab processing as related examples of HMW protein caused WRA (Boeniger et al. 2001; Ortega et al. 2002).

Intramural and extramural research support has been a major mechanism that underlies NIOSH activities and outputs relevant to diisocyanate asthma. Since 1996, RDRP intramural and NIOSH-funded extramural investigators have authored over 50 peer-reviewed publications covering toxicology, mechanisms of action, epidemiology, clinical aspects of disease and medical monitoring, sampling methods, and disease prevention and environmental controls (NIOSH 2006a). Highlights of this research output include the development of a tumor necrosis factor (TNF) receptor knockout mouse model that was used to demonstrate the role of TNF-α in diisocyanate sensitization, the first human study to confirm that dermal exposure increased the risk for pulmonary sensitization, and four published papers on 17 methods for diisocyanate exposure monitoring and analysis of samples. In addition to peer-reviewed publications, NIOSH analytic methods for diisocyanate quantification are available through the NIOSH website.

Between 1985 and 1994, NIOSH carried out 29 HHEs—identifiable by the key term search "isocyanate(s)" in the NIOSH database; between 1995 and 2004, the number fell to 11. It is not clear whether industry practices or other interventions account for this fall-off or whether other factors are responsible. In 1996, NIOSH published an important Alert, "Preventing Asthma and Death From Diisocyanate" (NIOSH 1996b). In 2006, a NIOSH Alert, "Preventing Asthma and Death from MDI [mineral-dust induced] Exposure During Spray-on Truck Bed Liner and Related Applications" (NIOSH 2006c), summarized four case reports: one death and several incidents of asthma or other respiratory disease that followed exposure to MDI during spray-on truck bed lining operations. In the same way that SENSOR data have allowed tracking of latex-related asthma, a new source of diisocyanate asthma has been detected by this surveillance system; the truck bed liner MDI cases were detected through the Michigan SENSOR project and reported in the MMWR.

NIOSH has disseminated information about diisocyanate asthma through other routes, including two major international conferences that featured presentations on diisocyanate-related research (on occupational asthma and skin exposures) that the agency sponsored. In addition, the RDRP developed a website that provides a single source of published references on the prevention of diisocyanate-induced asthma. Another noteworthy output was a memorandum of understanding between NIOSH and the American Chemistry Council Diisocyanates Panel. Signed in 2003, this agreement facilitates the identification and implementation of best practices in disease prevention and medical monitoring related to diisocyanates.

Although not fully described in the evidence package, NIOSH has conducted intramural research on potential mechanisms of diisocyanate-induced sensitization and asthma as well as mechanistic research on another class of LMW sensitizers, acid anhydrides (Zhang et al. 2002, 2004, 2006). In addition, an HHE of an outbreak of asthma due to another LMW sensitizer, 3-amino-5-mercapto-1,2,4-triazole, led to peer-reviewed publications, one of which documented sensitization in an animal model (Hnizdo and Sylvain 2003; Klink and Meade 2003; Hnizdo et al. 2004a). The RDRP has also played a key role in gathering, organizing, and interpreting research and case studies on the relationship between machining fluids and hypersensitivity pneumonitis as well as occupational asthma. This work included an important workshop (Kreiss and Cox-Ganser 1997), development of a criteria document that was used by a federal advisory committee established by OSHA (NIOSH 1998a, Metalworking Fluids Standards Advisory Committee 1999), and an assessment of metal-working fluid exposures in small workplaces that demonstrated these locations would not be adversely affected by NIOSH's recommended exposure limit (Piacitelli et al. 2001).

The "WRA in offices and schools" project has been the largest component of the RDRP Asthma Research Program Project (funding from 2000 projected through 2010). As described by NIOSH (2006a), the project involved field investigations that included clinical assessment of building occupants, development of biomarkers for improved assessment of exposure to bioaerosols, and a longitudinal study of the efficacy of remediation work in a damp building. Collaborations were established with the University of Sydney to design a halogen immunoassay to assess personal exposures to fungi, with the University of Cincinnati to develop immunoassay-compatible sampling techniques that make it possible to differentiate fungal spores from fungal fragments, and with the EPA to develop serologic immunoassays for biomonitoring exposures to fungi.

Since 1996, RDRP investigators supported through intramural and extramural funding mechanisms published multiple articles relevant to this topic, broadly defined, in the peer-reviewed literature. An RDRP investigator coauthored the relevant chapter in the latest edition of an influential textbook on occupational asthma (Menzies and Kriess 2006). Several monoclonal antibodies developed by RDRP investigators have been patented and licensed to companies for commercialization. An RDRP gas and vapor team established in 2001 to investigate the gas-phase and surface-phase chemistry of indoor environments has built an experimental laboratory to study volatile organic compounds and oxygen radicals. In collaboration with the Harvard School of Public Health, the RDRP cosponsored an international workshop on indoor chemistry and health in 2004. The RDRP also partnered with multiple federal agencies to organize the Surgeon General's Workshop on Healthy Indoor Environment held in 2005.

The scope of work on indoor air summarized above has the potential to lose "specificity" to the question of asthma alone. For example, indoor exposures to mold and endotoxin that are subsumed within this goal are associated with important respiratory tract responses other than asthma (as well as being linked to non-respiratory tract responses). Although building-related illness can include asthma, other health effects such as "sick building syndrome" have little direct relevance to the stated goal of preventing and reducing WRA related to nonindustrial indoor environmental quality. The potential for using resources theoretically targeted to WRA for studying non-asthma health effects is underscored by the record of HHEs in this area: only 19% of all indoor air investigations since 1982 appear to prominently feature asthma as a suspect complaint. Although not limited to asthma, the RDRP has published a systematic evaluation of data from office buildings for which HHEs were received in a 6-month period in 1992 and 1993 (Mendell et al. 2003).

Although the potential for loss of focus does exist with indoor air investigations, RDRP work on bioaerosols in office buildings and schools has relevance to occupational exposures to bioaerosols in other settings—for example, from contaminated metalworking fluids and in animal confinement facilities. Thus, on balance, although the indoor WRA asthma goal may include work on non-asthma-related issues, the benefits of this line of research outweigh this potential drawback.

To address the goal of improved WRA detection, NIOSH has taken into account both new onset of occupational asthma and aggravation of preexisting disease. There is some overlap with the subgoal of reducing WRA associated with indoor air quality in nonindustrial settings, to the extent that the latter is perceived as an emerging asthma problem. The WRA detection goal, however, is more broadly cast. A linchpin of the RDRP effort to improve the detection of WRA has been the SENSOR program. The SENSOR case definition for WRA is cited frequently in the published literature. The contribution of SENSOR to data on latex and diisocyanate asthma was discussed above.

In addition to standard surveillance reports from SENSOR, several key focused analyses based on pooled SENSOR data have been published in the peer-reviewed literature and include epidemiologic assessments of work-aggravated asthma and irritant-induced asthma (Henneberger et al. 2003a; Goe et al. 2004). Another investigation that built directly on SENSOR was a project to include questions on occupational asthma in the annual Behavioral Risk Factor Surveillance System survey when it was administered in three SENSOR states in 2001 (Breton et al. 2006; Flattery et al. 2006). The RDRP also led a key methodologic assessment of asthma surveillance that built on SENSOR data from one participating state and applied an innovative capture-recapture strategy to better estimate the true incidence of occupational asthma (Henneberger et al. 1999). With a targeted request for application collaborative agreement strategy, the RDRP also has supported analyses of several different health care databases to better estimate the incidence of new onset occupational asthma as well as work-aggravated asthma (Henneberger et al. 2003a,b; Vollmer et al. 2005; Henneberger et al. 2006; Sama et al. 2006).

The RDRP played a key role in facilitating the ATS Statement on the Occupational Contribution to the Burden of Obstructive Airways Disease (Balmes et al. 2003), a frequently cited document that summarized the large body of literature (through 2000) on the work-relatedness of a substantial fraction of cases of adult asthma. The partnership between the RDRP and the ATS grew out of the NORA asthma and COPD team deliberations. This document and the RDRP-ATS partnership are discussed below in the subsection on fixed obstructive airway disease. Beyond the activities and outputs specifically related to the subgoal of improved detection of WRA summarized above, the RDRP partnered with the ATS to develop a survey instrument to assess respiratory symptoms in adults that included an occupational module; the RDRP cosponsored a meeting of the committee for further development of the questionnaire. After a draft questionnaire was completed, the RDRP contracted with the NCHS to conduct cognitive testing of the new core and occupational questions. NIOSH (2006a) stated that cognitive testing of the revised adult respiratory questionnaire will be completed in 2007 and data from that testing will be used in 2008 to modify parts of the questionnaire related to occupational exposures, symptoms, and diseases. The RDRP worked to develop hardware (a "belt spirometer") and software to facilitate the collection of serial spirometric data for use in detecting WRA; some of this work was done in collaboration with the manufacturer of a popular portable spirometer, the EasyOne (ndd Medizintechnik AG).

Assessment of Relevance

Given the high prevalence of new cases of adult asthma that may be related to occupational exposures, the four WRA subgoals are highly relevant to improved occupational safety and health among the U.S. workforce. As presented by NIOSH (2006a, p. 116),

> WRA is the most common respiratory disease treated in occupational health clinics in the U.S., accounting for a substantial share of all asthma among adults. A 2003 statement of the American Thoracic Society indicates that 15 percent of asthma among adults is attributable to work [Balmes et al. 2003]. Under cooperative agreements, NIOSH-funded researchers have since estimated that 29 percent to 33 percent of new-onset adult asthma is attributable to work. In addition, RDRP scientists estimate that 23 percent of existing adult asthma is exacerbated by work. Assuming that about 25 percent of asthma is attributable to work, at least 2.25 million Americans between the ages of 15 and 65 have experienced onset or exacerbation of their asthma due to workplace conditions [Sama et al. 2006].

The planning, activities, and outputs described above are judged to be of generally high quality for each of the four subgoals. The work on latex represents a timely response to an epidemic of occupational asthma and a model of how effective planning can lead to research targeted to generate knowledge that can be transferred to improve outcomes. The work on diisocyanates directly addressed the most common LMW sensitizing cause of asthma in the developed world and again led to important knowledge that has been used to prevent disease. While the work of the RDRP on indoor environmental quality is judged to be relevant to occupational health and safety in the general sense, it is not always necessarily related to WRA. Some of the work on indoor environmental quality has a broader relevance (e.g., to residential settings). The RDRP effort to improve the detection of WRA in the broad sense is of the highest relevance. Increased awareness of this disease has been engendered by RDRP's support of the ATS statement, analysis of National Health and Nutrition Examination Survey (NHANES) data, and population-based research conducted through collaborative agreements.

Assessment of Impact

RDRP's work to achieve the four WRA subgoals has had considerable impact. In the case of latex, the RDRP investigators have documented a fall in the prevalence of latex sensitization as a result of the intervention effort begun with the 1996 NIOSH Alert. For diisocyanates, RDRP's work has contributed to greater awareness of WRA due to this class of LMW sensitizers. This work included identifying a new source of exposure, with publication of a report in the MMWR, and a new NIOSH Alert in response. With regard to WRA and indoor air quality, the greatest impact of RDRP research appears to be in the development of methods for improved quantitative assessment of exposure to fungi. The impact of RDRP efforts to improve detection of WRA in the broad sense is more difficult to assess given the lack of adequate surveillance data, but the quality of transfer activities in this area (e.g., the ATS statement on the "Occupational Contribution to the Burden of Obstructive Airway Disease") is judged to be high.

Fixed Obstructive Airway Diseases

The four subgoals for fixed obstructive airway diseases formulated by NIOSH are as follows:

- Establish the work-relatedness of COPD.
- Develop tools and identify at-risk workers in industries and occupations to assess the extent, severity, and burden of work-related COPD.
- Develop, test, and disseminate recommendations for preventing COPD in the workplace.
- Prevent and reduce flavoring related bronchiolitis obliterans.

The first three subgoals are quite broadly defined. This is appropriate, given the current state of the art in the epidemiology of work-related COPD. Moreover, an overly specific definition of COPD (e.g., based solely on a decreased FEV_1/FVC ratio, where FEV_1 is forced expiratory volume in the first second and FVC is forced vital capacity) could lead to the lack of consideration of certain related conditions. For example, chronic bronchitis is frequently addressed as a condition captured under the rubric of COPD. The term COPD typically is used in a way intended to include emphysema as well. Although the NIOSH goals did not explicitly endorse a definition of COPD as a subsuming disease defined by air flow obstruction, emphysema (defined by loss of air-exchanging units of the lung and supporting structure), and chronic bronchitis (defined by chronic sputum production), it was implicit in the supporting materials provided. By the same token, even though "prevention" is not specifically defined in the goal, elements of secondary and tertiary as well as primary prevention are relevant. For example, efforts can be made to reduce work-related disability associated with COPD.

Because of the leading role of the RDRP in the research effort to understand, control, and prevent bronchiolitis obliterans in the flavoring industry, the fourth subgoal also is consistent with the priorities of the RDRP. Although it could be construed to be cast narrowly, especially juxtaposed with the prior three subgoals, such an interpretation would overlook both the inherent importance of this disease outbreak in its own right and the application of this work to other established or emerging occupational airway disease processes characterized more by fixed than by reversible airway obstruction. Such conditions (including byssinosis and irritant inhalation-related bronchiolitis obliterans) typically are considered to be separate from WRA. By establishing a subgoal focused on flavoring-related bronchiolitis obliterans, NIOSH provides for a mechanism to prioritize research in this general area and in so doing addresses a topic that has been understudied and overlooked from a preventive standpoint.

A final point to emphasize in assessing the subgoals that target work-related COPD is that the classic pneumoconioses, although leading to a restrictive ventilatory deficit, also can be associated with COPD, chronic bronchitis, and emphysema. This is most clearly the case with coal workers' pneumoconiosis (CWP). Although RDRP research related to CWP and other fibrotic (restrictive) lung diseases are addressed separately, activities that address airway effects of exposures to fibrotic agents are discussed in this section.

Planning and Production Inputs

As noted above, a partnership between the RDRP and the ATS grew out of the NORA asthma and COPD team deliberations and led to the establishment of a committee that undertook a systematic review of the published literature on the question of work relatedness of COPD as well as of asthma. Not only did the RDRP provide logistic support for the work of this committee, but three RDRP investigators were key participants directly involved in writing the report that was the committee's main output (see below).

Both before and in follow-up to publication of the ATS statement, the RDRP has approached the need for better assessment of the magnitude of the burden of work-related COPD on an industry-specific and occupation-specific basis.

This includes conducting a series of epidemiologic studies, developing improved methods for identifying COPD, and collaborating with extramural researchers at various universities. For example, the National Study of Coal Workers' Pneumoconiosis (NSCWP) included spirometry data relevant to obstruction; in follow-up to initial observations, RDRP investigators planned a series of thoughtful research investigations to answer specific questions related to coal-dust-induced COPD. Another notable example of relatively new RDRP activity that was not industrial sector based was the use of NHANES data to address the population-attributable risk of COPD due to occupational exposures and to provide support for greater preventive efforts in this area.

The RDRP has approached the prevention of work-related COPD primarily through research on spirometric monitoring of lung function at the workplace. One aspect of this approach involves spirometry as a medical surveillance tool, consistent with NIOSH's recommendations for a variety of exposures. From a production input perspective, RDRP efforts to develop standard population reference equations for predicted lung function values represent a crucial contribution to preventing work-related COPD. First, these equations (derived from NHANES data collected following the design by and supervision of NIOSH research scientists) allow for race/ethnicity-specific estimations for Caucasians, Blacks, and Hispanics. Second, the equations facilitate the identification of exposure-associated lung function deficits within a cohort based on a reliable external comparison group, including, if needed, an external smoking adjustment parameter estimate. This RDRP effort also provided NCHS with prediction equations for use in settings other than the workplace, which is recognized as a highly valuable contribution.

The RDRP research effort on flavoring-induced bronchiolitis obliterans began with a 2000 HHE request from the Missouri Department of Health. The results of the initial RDRP investigation led to the recognition that workers throughout the flavoring industry also might be at risk for this potentially debilitating airway disease. A multidisciplinary approach to the problem was organized and involved epidemiologists, inhalational toxicologists, industrial hygienists, and engineers. This multidisciplinary approach is similar to that used in the RDRP response to the initial report of nylon flock workers' lung disease (see interstitial lung diseases below). The RDRP approach to these two clusters of cases of potentially severe occupational respiratory disease has been of great value in defining the scope of the threat to worker health.

The response to the HHE request with regard to a cluster of cases of bronchiolitis obliterans from a single microwave popcorn manufacturing facility can be considered a model of how an HHE (a surveillance and initial data-gathering instrument) should serve as a key production input for setting priorities and planning research objectives relevant to detecting and preventing occupational diseases. An earlier HHE in 1985 investigated a cluster of cases of bronchiolitis obliterans in a commercial baking facility that used diacetyl (McConnell and Hartle 1986). The failure of the earlier HHE to trigger a deeper investigation that might have highlighted the hazards of diacetyl further underscores the point that, without a concerted effort to make fixed obstructive airway disease an RDRP priority, important surveillance "signals" can and will be lost. In follow-up planning efforts intended to better define the population of workers at risk for flavoring-induced bronchiolitis obliterans, RDRP investigators have attempted to engage the Flavor and Extract Manufacturing Association in industry-wide surveillance activities, have held workshops and exchanged information with specific industries, and have responded to requests for assistance from the California OSHA and Department of Health Services. Beyond the specific case of diacetyl, the more general issue of the systematic ways the RDRP could look for "signals" within the HHE program and other surveillance resources should be considered in future planning.

Activities, Outputs, and Outcomes

RDRP activities and outputs on coal-dust-induced COPD derived from the NSCWP have been considerable and comprehensive. The exposure-response relationship was investigated initially by relating quantitative measures of dust exposure to measurements of pulmonary function obtained serially in the NSCWP (Attfield and Hodous 1992). These and related data helped support an updated NIOSH criteria document for coal mine dust published in 1995. More recently, the RDRP has focused on identifying occupational and nonoccupational risk factors for excessive declines in pulmonary function among coal workers and the implications of these risk factors for morbidity and mortality (Hodgins et al. 1998; Wang et al. 2005). Additional data were collected to supplement the NSCWP for these studies. Laboratory studies that included pathological evaluation of material collected through the National Coal Workers' Autopsy Program and toxicologic investigations that involved potential pathways by which silica can induce lung injury by generating reactive oxygen species contributed to a mechanistic understanding of coal-dust-induced COPD (Hnizdo and Vallyathan 2003).

Other relevant RDRP research has included investigations of the association between excessive decline in FEV_1 and retirement due to chest illness and the exposure-response relationship for mortality due to chronic bronchitis or emphysema (Kuempel et al. 1995; Beeckman et al. 2001). Combined with the pathological findings from the autopsy study noted above, these results indicate that exposure to coal dust can lead to permanent and debilitating structural damage to the airways and lungs. The RDRP also has supported laboratory-based investigations related to coal mining and airway disease (Cohen et al. 2002).

While past RDRP epidemiologic research on work-related COPD extended beyond coal mining, the work was limited to industry-specific exposures or discrete occupations and was not framed to address more general cross-industry or cross-occupation exposures to vapors, gas, dust, and fumes. Cotton dust exposure, which is linked to fixed obstructive airway disease as a manifestation of Stage IV byssinosis, is an excellent example of such a non-coal-industry-specific research focus. NIOSH has provided the key extramural funding for the longitudinal study of cotton dust exposure and fixed airway obstruction among textile workers in China, establishing one of the longest-followed byssinosis cohorts. This study has led to improved knowledge of both exposure-response and individual risk factors. NIOSH also has provided extramural support for studies of grain-dust-related and diesel-exhaust-related respiratory health effects that are highly relevant to occupationally related COPD. In addition, NIOSH's support of the Agricultural Research Centers has helped to promote research relevant to organic-dust-related obstruction—for example, research on the mechanisms underlying chronic bronchitis in farmers (Lambert et al. 2005).

Although industry- and occupation-specific studies have provided key early evidence of the link between work-related exposures and COPD, more recently, analyses of existing data from large population-based surveys have been used to provide a more global picture of the impact of occupational exposures on the prevalence of COPD in the general population. NIOSH has supported this work through extramural support or it has been conducted internally by RDRP investigators. Extramural funding has supported collaborations with several universities (University of California, Los Angeles; University of Oregon; Tulane University) that provided for analysis of the effects of occupational factors on the risk of COPD in cohorts to which RDRP investigators otherwise would not have had access (e.g., the Lung Health Study, Kaiser Permanente-Northwest Region, longitudinal spirometric data from 12,000 workers in 11 industrial facilities). The results of the collaborative analyses have not been published other than as abstracts, but the approaches taken (e.g., use of a new job-exposure matrix) should provide valuable contributions. Another important output was a peer-reviewed paper on the costs associated with work-related COPD, which was a collaborative effort between University of California, Davis, investigators and an RDRP investigator (Leigh et al. 2002). The results of the intramural analyses of NHANES III data have been published in two peer-reviewed papers and are a major contribution to the scientific literature on COPD (Hnizdo et al. 2002, 2004b).

An output with wide distribution was the ATS Policy Statement on the "Occupational Contribution to the Burden of Obstructive Airway Disease" (Balmes et al. 2003) to which the RDRP made a major contribution. The RDRP cosponsored a meeting of the international writing committee in Morgantown in 2001 that greatly facilitated the completion of this document.

The RDRP development of new reference equations for spirometry, discussed previously (Hankinson et al. 1999), has been an extremely valuable contribution to the improved detection of COPD. Developed as a consequence of the RDRP's responsibility for spirometry in NHANES surveys, these equations have become the "gold standard" in the United States. The RDRP also has activities directed toward improving lung function testing of workers, including hardware and software development, methodologic work on analysis of longitudinal decline and identification of individuals with excessive declines, and evaluation of decision rules based on excessive decline. Since 2000, this work, in partial collaboration with Tulane University, has led to multiple papers published in the peer-reviewed literature and represents an impressive body of work (Hnizdo et al. 2005, 2006a). Another recent paper by RDRP investigators demonstrates how different lung-function-based case definitions for COPD can influence prevalence data (Hnizdo et al. 2006b). In 2006, the RDRP also organized a workshop that reviewed the current statistical methods for longitudinal spirometry assessment.

The RDRP has developed sampling methods to measure exposure to volatile chemicals, including diacetyl in facilities that process butter flavoring. A longitudinal investigation of the sentinel plant in Missouri and new investigations at five additional microwave popcorn plants have been conducted (Kreiss et al. 2002; Kanwal et al. 2006). These studies showed that workers were at risk throughout the industry and that the highest risk appears to be associated with peak exposures to butter flavoring chemicals due to open handling, even when ventilation maintained low average exposures. Animal inhalational toxicology studies conducted by the RDRP have demonstrated that diacetyl can injure airways, although other ingredients in butter flavoring may contribute to toxicity (Hubbs et al. 2002; Fedan et al. 2006). RDRP investigators also performed laboratory analyses of emissions from bulk butter flavoring samples collected at six microwave plants and demonstrated that emissions of volatile chemicals were greater from liquids and pastes than from powders (Boylstein et al. 2006). To better assess the potential for flavoring-related lung disease in other food-production facilities, RDRP investigators have done walk-through surveys at plants that produce beverages, flavored coffee, and packaged oil. Presentations to the Flavor and Extract Manufacturers Association have been made in an effort to establish partnerships with member companies to work with the RDRP to assess worker risk throughout the industry. RDRP investigators also have partnered with the California OSHA and Department of Health Services to conduct surveillance for flavoring-related respiratory disease and to help develop a state-based standard for California, an ongoing process (MMWR 2007b).

Assessment of Relevance

The four fixed obstructive airway disease subgoals are highly relevant to improved occupational safety and health among the U.S. workforce in light of evidence that up to 15% of all cases of COPD may be related to occupational exposures (Balmes et al. 2003; Blanc and Torén 2007). Further, as described by NIOSH (2006a, p. 139):

> COPD is the fourth leading cause of death in the U.S. In 2003, 10.7 million U.S. adults were estimated to have COPD, although close to 24 million adults had evidence of impaired lung function, indicating under-diagnosis of COPD in the U.S. The cost to the nation of COPD was approximately $37.2 billion in 2004, including healthcare expenditures of $20.9 billion in direct costs, $7.4 billion in indirect morbidity costs, and $8.9 billion in indirect mortality costs [American Lung Association 2006].

The planning, activities, and outputs described above are judged to be of high quality for each of the four subgoals. RDRP efforts that included the ATS statement, industry-specific studies (especially those in coal mining), and population-based studies directly addressed underrecognition of the work relatedness of COPD. Work on spirometric methods, as well as in industry-specific and population-based studies, also directly addressed the identification of at-risk workers and assessment of the burden of work-related COPD. Much of the work on spirometry, especially the development of equations to predict normal values, has an even broader relevance for preventing COPD. The work on flavoring-related bronchiolitis obliterans has been a well-coordinated and integrated multidisciplinary response to an emerging problem that can be considered a model for future efforts of the RDRP.

The need for improved surveillance data on work-related COPD is critical for assessing relevance. Without such data, it is difficult to appropriately allocate limited resources for exposure and medical monitoring and interventions to prevent and control disease. The RDRP plan to collaborate with the NCHS and the National Heart, Lung, and Blood Institute to provide support for spirometry in the 2007-2008 NHANES is of highest importance; failure to do so would be a major setback for COPD surveillance at the national level.

The addition of new questions on occupational exposure to the NHANES survey is a major improvement; analysis of NHANES on the basis of occupational and industrial categories, which can be segregated with reasonable precision, has already yielded important results, as summarized previously (Hnizdo et al. 2004a). However, inadequate geographic resolution of NHANES data prevents appropriate analysis of state-level variation in COPD outcomes as opposed to large regional aggregations (e.g., the western United States). Therefore, RDRP efforts to seek new methods for conducting work related COPD surveillance are of great importance for understanding the relevance of COPD-related research and appropriately targeting resources.

Assessment of Impact

RDRP work to achieve the four fixed obstructive airway disease subgoals has had a considerable impact. Although the ATS statement on the "Occupational Contribution to the Burden of Obstructive Airways Disease" was mentioned previously in relation to WRA, its impact on increased recognition of work-related COPD should not be understated. Multiple peer-reviewed papers by RDRP investigators report the results of studies using NSCWP data and provide clear evidence that certain groups of U.S. underground coal miners are at risk for COPD, pneumoconiosis, bronchitis, and emphysema, which are associated with disability and mortality. Numerous investigators and policy makers have cited this body of research in their assessment of the links between respirable dust exposure, diesel particulate matter, and COPD, not only for this specific industry but also as a measure of biologic plausibility for COPD related to other exposures. RDRP studies using NHANES data have contributed to a greater recognition of the role of occupational factors in the burden of COPD on the U.S. population. RDRP work on spirometry, especially the development of new reference equations for normative values, has had a major impact on respiratory disease research in general and, more specifically, on preventing COPD. The RDRP response to the initial outbreak of diacetyl-induced bronchiolitis obliterans has led to surveillance efforts in multiple locations in an effort to detect and prevent disease. In large measure due to RDRP research, the California OSHA is considering establishment of a permissible exposure limit (PEL) for diacetyl.

STRATEGIC GOAL 2: PREVENT AND REDUCE WORK-RELATED INTERSTITIAL LUNG DISEASES

Introduction

Historically, research and prevention activities, outputs, and outcomes that pertain to work-related interstitial lung diseases generally categorized as pneumoconioses have been a critical part of the RDRP mission. Because of the importance of CWP, initial RDRP research focused on it almost entirely. A number of research, surveillance, and regulatory-related activities associated with interstitial lung diseases first mandated by the Federal Mine Health and Safety Act of 1969 were later assigned to NIOSH. Beginning in 1976, with formation of the Division of Respiratory Disease Studies, the scope of respiratory disease research extended significantly to other forms of interstitial lung disease, including silicosis, chronic beryllium disease (CBD), and asbestosis. In addition, work on organic dusts has subsumed hypersensitivity pneumonitis (also known as extrinsic allergic alveolitis), another potentially fibrotic lung condition. More recently, novel emerging occupational lung conditions have been recognized that also have prominent fibrotic features—for example "flock worker's lung." The following sections summarize the activities of the RDRP that pertain to interstitial lung diseases.

The four subgoals related to the prevention and reduction of work-related interstitial disease as formulated by the RDRP are as follows:

- Prevent and reduce respiratory diseases induced by coal mine dust.
- Prevent and reduce silica-induced respiratory diseases.
- Prevent and reduce fiber-induced diseases.
- Prevent and reduce CBD.

The first of these subgoals clearly remains central to core RDRP activities. Although coal mine dust had long been known to cause pneumoconiosis and chronic respiratory disease, little was done in the United States to prevent CWP until passage of the Federal Mine Health and Safety Act of 1969, as alluded to above. This legislation established the federal standard of 2 mg/m^3 for respirable coal mine dust; before this standard was established, between 150,000 and 500,000 U.S. underground miners were typically exposed to respirable dust concentrations of 6-8 mg/m^3 (Attfield and Sexias 1995). Moreover, new data are emerging that suggest geographic "hot spots" within certain states and even across multistate regions are linked to endemic CWP characterized by rapid progression of disease (Antao et al. 2005; MMWR 2006a, 2007a). Moreover, emerging NIOSH surveillance data indicate that the prevalence of CWP among workers with 25 years or more of exposure is climbing relative to the historic lows in the late 1990s (NIOSH 2006a; Ward 2007) (see further discussion below). Finally, this specific subgoal, as articulated, overlaps with the RDRP initiative on work-related COPD (addressed in the preceding section). Thus, there is more than sufficient rationale for the coal-dust-related subgoal.

Silicosis, the pneumoconiosis associated with silica exposure, has been a long-recognized and continuing problem in many industrial sectors in the United States. Unlike CWP, which by definition is a mining-specific issue, silica exposure in mining and quarrying accounts for only one of multiple potential exposure venues. Traditional sources of exposure include heavy industry (such as foundry work, glass making, and pottery manufacture) as well as construction trades in which sandblasting remains an ongoing risk factor both for those directly carrying out the work and for "bystander employees." Not only do silica-containing blasting materials remain in use, but there is growing awareness (in part through research work supported by extramural NIOSH programs) that concrete finishing operations that have become a prominent component of construction can be associated with substantive silica exposure (Woskie et al. 2002; Croteau et al. 2004; Akbar-Khanzadeh et al. 2007). Beyond the well-defined risk of silicosis, other silica-associated adverse health effects are relevant to this subgoal, although they are not limited to interstitial processes. These effects include "acute silicosis" (which overlaps with pulmonary alveolar proteinosis), silica-promoted mycobacterial infection (including atypical mycobacterial disease), silica-associated extrapulmonary rheumatologic disorders, and, overlapping with RDRP goals in respiratory cancer, the association between silica and bronchogenic carcinoma. Considering all these disease links, coupled with ongoing sources of exposure, this subgoal has unquestionable relevancy.

The subgoal related to "fiber-induced" lung disease as formulated subsumes both inorganic and organic (naturally occurring and anthropogenic) materials that, because of their physical characteristics, may share important mechanistic pathways and may benefit from overlapping prevention strategies. Asbestos is by far the most prominent of these exposures in terms of past commercial applications and ongoing and likely future respiratory morbidity and mortality. Other natural mineral fibers closely resemble commercial asbestos and also cause interstitial lung disease (e.g., amphibole fibers that contaminate vermiculite ore once mined in Libby, Montana). Other naturally occurring mineral fibers (including fibrous talc, wollastonite, attapulgite, and mordenite fibers) differ from asbestos and amphibole-contaminated vermiculite minerologically and toxicologically but nonetheless may be relevant to this subgoal.

Synthetic vitreous (inorganic) fibers, including refractory ceramic fibers (RCFs), increasingly used as asbestos substitutes may be hazardous, especially if they are durable and have dimensional characteristics mimicking asbestos. Fibrous glass and mineral wool are examples of synthetic vitreous fibers that may be less durable but still carry lung health risks. Modern synthetic organic fibers (e.g., nylon, rayon, polyester, and others), depending on their processing and applications, pose a respirable fiber hazard and the potential for interstitial lung disease, of which the emerging disease flock worker's lung appears to be prototypical.

While the total population at risk for all the various fibers that may be subsumed under this subgoal has not been determined, it is estimated that nearly 4 million workers are at risk to a subset of these hazardous fibers (asbestos, fibrous talc, fibrous glass, wollastonite, attapulgite fibers, and RCFs) (NIOSH 1986b, 2006c). These exposure data, although fragmentary, underscore the importance of this subgoal.

Beryllium, the focus of the fourth subgoal, is a lightweight metal with properties that include heat resistance and conductance, electrical conductance, flexibility, formability, neutron moderation, x-ray transparency, and lubricity. These properties have led to beryllium being used in many applications in a variety of settings, including the nuclear industry, aerospace and electronics/microelectronics manufacturing, and a variety of specialty applications, including health care. As the uses for beryllium have become increasingly diverse, exposures are more widespread. It is estimated that as many as 134,000 current U.S. workers have been or will be exposed to beryllium at some time during their working lives, which may be an all time high (Henneberger et al. 2004).

Exposure to beryllium can lead to sensitization, resulting in CBD, a granulomatous, often progressive, potentially fatal, fibrotic lung disease that is histologically identical to the idiopathic disease sarcoidosis. Factors leading to sensitization and then to disease among sensitized workers have been only partially elucidated. NIOSH-supported research efforts have been pivotal in this work (see below). Through this work, it has become clear that very low levels of exposure can carry substantial risk (Henneberger et al. 2001; Day et al. 2007) and clusters of disease in work settings not previously considered at risk have been well documented in recent investigations (Sackett et al. 2004; Welch et al. 2004). The relevance of beryllium sensitization and CBD goes beyond the population directly at risk, large and growing as it may be, because the causes and prevention of this process have relevance to additional, albeit less common, metals-associated, work-related granulomatous processes (e.g., disease caused by titanium and zirconium). Further, the beryllium subgoal may also be applicable to more general epidemiologic efforts to study potential occupational associations with sarcoidosis. For example, an otherwise unexplained increased incidence of sarcoidosis has been observed in dust-exposed firefighters in follow-up to the World Trade Center disaster (Izbicki et al. 2007).

The four subgoals, as formulated, are not as inclusive as they might be insofar as certain occupationally related fibrotic lung disease is concerned. For example, the previously cited group of conditions known as extrinsic alveolitis does not easily fall under any of the four stated subgoals. Organic dusts (as opposed to fibers) altogether seem to fall outside this particular set of subgoals. So too do gases or fumes that might lead to interstitial lung processes—for example, chemically induced bronchiolitis obliterans organizing pneumonia. We recognize that these omissions are largely an artifact of the structure of this review process rather than reflecting a fundamental lack of programmatic content on the part of the RDRP.

The following sections summarize planning and production inputs; activities, outputs, and outcomes; and assessment of relevance and impact, in turn, for each of the subgoals discussed above. Where appropriate, this discussion is extended to other selected RDRP initiatives that are relevant but may not otherwise be easily categorized within this format.

Respiratory Disease Induced by Coal Mine Dust

Planning and Production Inputs

The work of the RDRP addresses three primary objectives directed at disease assessment and prevention of occupational lung diseases among coal miners: research into disease causation, disease surveillance and monitoring, and research to improve measurement and control of coal mine dust. The three major subgoals of the program translate into the following:

- Verification that the currently enacted coal mine dust standard protects mine workers.
- Tracking disease occurrence in coal miners to document the status of prevention and to target preventive efforts.
- Improved measurement and controls to reduce exposure to coal mine dust.

Since the late 1960s, the coal industry has contracted to an estimated 74,000 people, mainly men, of whom 57% are underground coal miners. CWP has been documented to occur among both underground and surface coal miners. The legacy of exposure to respirable coal mine dust has been the well-documented impairment from advanced pneumoconiosis and airway obstruction that results in significant morbidity and accompanying compensation costs. A primary planning input is tracking the occurrence of disease in coal miners that was mandated in the original Federal Coal Mine Health and Safety Act. To fulfill the mandate, the RDRP operates x-ray surveillance programs to identify those with pneumoconiosis. RDRP efforts to raise employer and miner awareness has increased employer x-ray surveillance program compliance from 90% to 98% (2003-2006) and has increased miner participation from 20% to 30% (1985-1999) to nearly 50% (2003-2006). NIOSH realized very early in its surveillance program that evidence of CWP suffered from variability in the classification of x-ray evidence, which led NIOSH to create the B Reader Certification Program to standardize and test physicians' competency in CWP radiographic assessment. NIOSH has revised its B Reader Certification Program to comply with changes in the pneumoconiosis classification system of the International Labour Office (ILO). This includes collaborating in a multinational film reading trial led by RDRP scientists, which provided the basis for the ILO revision in collaboration with the American College of Radiology (NIOSH 2006a). The RDRP has held numerous workshops and major meetings that involve the

B Readers Certification Program. In addition to a continuing emphasis on CWP surveillance and associated methodology, the RDRP has successfully incorporated CWP morbidity and mortality in their series of WoRLD Surveillance Reports that have been widely disseminated nationally and internationally (see WoRLD Surveillance Reports 2002 [NIOSH 2003] and the eWoRLD Surveillance System, a more frequently updated web-formatted reporting system [NIOSH 2007a]).

Despite significant progress in reducing CWP, surveillance reports have identified rapidly progressive CWP in a geographic area that includes Eastern Kentucky and Southern West Virginia (Antao et al. 2005; MMWR 2006a, 2007a). NIOSH (2006a) stated "The reasons for this apparent rapid progression and continuing occurrence of progressive massive fibrosis are unclear. It may result from a combination of inadequacies in the present dust limit and its method of enforcement. These occurrences point to the need for continued surveillance and monitoring activities to track disease decline and to target enhanced exposure assessment and dust control." The RDRP has responded by conducting special radiographic examinations (Miners' Choice Program) from 1999 through 2002. The RDRP is now operating a mobile examination unit and is working with MSHA to determine hot spot surveillance that is supplemented by intense publicity and outreach efforts to enhance miner participation in these areas.

Activities, Outputs, and Outcomes

Research into disease causation includes assessment and verification that the enacted coal mine dust standard protects miners. The standard of 2 mg/m^3 was derived from British epidemiologic research, but few of the assumptions that underlie the 2 mg/m^3 PEL have since been validated among U.S. coal miners. RDRP efforts to validate the respiratory dust standard for coal miners included a series of comprehensive cross-sectional epidemiologic surveys, the National Coal Workers' Pneumoconiosis Study, that used standardized methodologies for chest radiography and spirometry to establish the prevalence and severity of disease in the United States; requirements that employers offer chest radiographic surveillance to all underground miners in the United States and that the RDRP assist in developing standardized radiographic reading methods and programs; and epidemiologic and laboratory research to explore, examine, and evaluate knowledge and assumptions about mechanisms and progression of disease among U.S. coal miners.

Led first by the Appalachian Laboratory for Occupational Respiratory Diseases, which then became the Division of Respiratory Disease Studies, the RDRP published more than 220 epidemiologic studies, surveillance laboratory-based studies, and dust control reports that documented a decreasing prevalence of CWP, declining severity of associated respiratory impairment, and improved dust control and monitoring measures. The completion of more than 300,000 radiographic surveillance examinations, supplemented by the serial National Coal Workers' Pneumoconiosis Study, provided reliable U.S. occupational dose-response data for CWP that resulted in a NIOSH criteria document, "Occupational Exposure to Respirable Coal Mine Dust" (NIOSH 1995). That document recommended a respirable coal mine dust PEL of 1 mg/m^3 (NIOSH 1995, 2006a), which MSHA did not adopt.[1] These recommendations arise from a series of RDRP peer-reviewed publications (Attfield and Morring 1992a,b). With issuance of the criteria document in 1995, the focus of RDRP fieldwork has moved toward surveillance and prevention activities.

To comply with the respirable coal mine dust PEL set in the Federal Mine Safety and Health Act of 1969, the Bureau of Mines was required to improve dust exposure assessments, develop cleaner coal-cutting technology, and improve ventilation and dust suppression methods. This research included developing real-time dust monitors. Between 1996 and 2005, 97 publications on dust measurement and control were published by respiratory disease and mining research programs. Many of these publications are technical reports directed to the mining industry and to MSHA (Campoli et al. 1996; Cantrell et al. 1996; Belle et al. 2000; Bugarski and Gautam 2001). Development of these technologies was critical, as average production of continuous miner sections more than doubled between 1971 and 2003, and average long-wall production increased nearly 10-fold during the same period.

Assessment of Relevance

Current program efforts for RDRP research to further reduce pneumoconiosis among coal miners have developed the following objectives:

- Improve technologies for dust assessment and dust control.
- Monitor and evaluate the extent, severity, and characteristics of CWP.
- Increase awareness of the occupational contribution to pulmonary impairment associated with CWP.
- Research, document, and better understand rapidly progressive CWP.
- Research, document, and recommend targeted dust-reduction strategies, digital chest radiographs, and improved radiographic assessment methods for CWP.

All these activities are highly relevant for reducing the incidence of CWP.

The need to improve technologies to assess respirable dust and to monitor levels of respirable dust is critical to reducing the incidence and severity of CWP.

---

[1]The reasons underlying MSHA's decision were not presented to this committee for consideration.

The experience gained in the CWP program creates a highly valuable framework for better understanding and preventing the recently described rapidly progressive CWP. Efforts to understand the strengths and limitations of digital radiographs in detecting pneumoconioses and for implementing the use of digital radiography in the B Reader Certification Program are important.

Assessment of Impact

The impact of NIOSH's work quantifying the cumulative incidence of respiratory diseases induced by coal mine dust has been very high. Unfortunately, it has not solely been an exercise in documenting an ongoing reduction in the burden of disease, given the findings of hot spots of rapidly progressive CWP and the apparent increase in the prevalence of CWP across the United States (MMWR 2006a, 2007a; Ward 2007). The RDRP and the Mining Research Program have collaborated to develop, disseminate, and significantly improve control of coal mine dust in underground mines. For continuous miner work, dust samples exceeding the PEL of 2 mg/m^3 have dropped from 49% in 1971 to 9% in 2003. For longwall miners, the percentage dropped from 44% to 12% over the same time period (NIOSH 2006a). Data provided to this committee (NIOSH 2006a) indicate that, between 1970 and 1999, the percentage of all underground coal miners examined with CWP (ILO category 1/0+) declined from 11% to 1.6%. As shown in Figure 2-2, for miners with 25 years of tenure or longer, the prevalence of CWP dropped from 35% (1983-1978) to below 5% (1997-1999) and to just over 5% in 2004 (NIOSH 2003). However, Figure 2-2 also indicates an upswing in the disease in 2004-2005. More recent data indicate further increases in the prevalence of CWP. Data from 2005 and 2006 indicate that about 9% of miners with tenure of 25 years or longer showed lung abnormalities indicating CWP. Rates among miners with 20 to 24 years of experience increased from 2.5% to about 6% over the same period (data from E. L. Petsonk, NIOSH, as cited by Ward [2007]). The geographic hot spots of cases of rapidly progressive CWP described above are also of great concern. As described by Antao et al. (2005), "cases of rapidly progressive CWP can be regarded as sentinel health events, indicating inadequate prevention measures in specific regions."

The x-ray program has resulted in 3,000 of 18,000 eligible miners with radiographic evidence of pneumoconiosis exercising their right to transfer to a less-dusty job. This also means that 15,000 have not transferred to jobs with reduced exposure. Moreover, as stated above, one in two miners does not even participate in the program.

RDRP activities affect other issues related to lung disease in coal workers that are substantive NIOSH outputs, most notably for occupationally related COPD (see

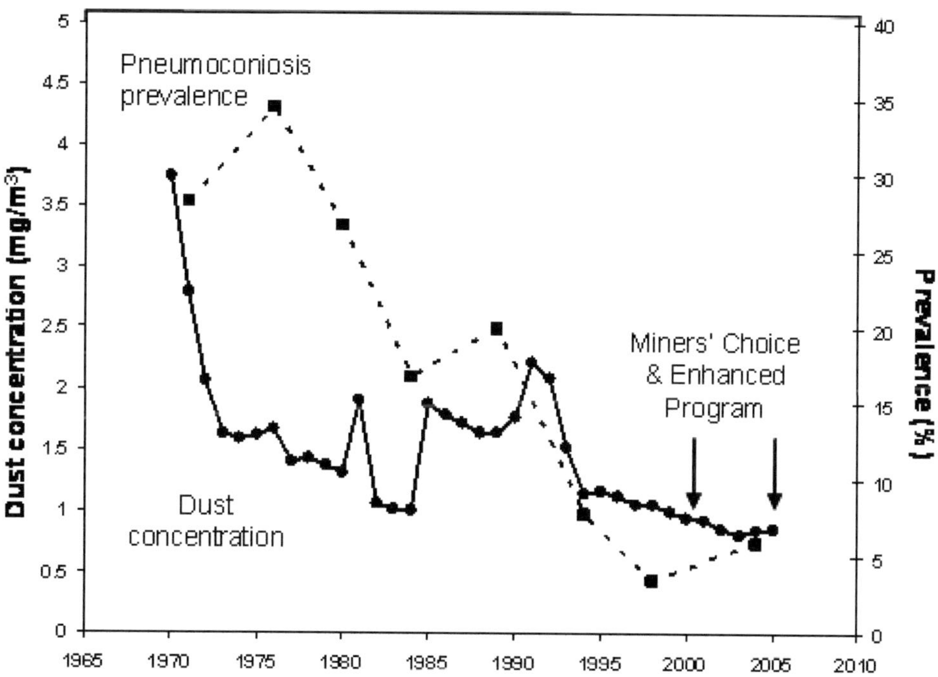

FIGURE 2-2 Reductions in respirable dust concentrations derived from MSHA compliance data, and prevalence of CWP among participants in the Coal Workers' Health Surveillance Program, 1970-2005. Source: NIOSH 2006a.

above). Work on isocyanate-induced asthma is noteworthy, given the application of such materials in mining (NIOSH 1994; Donovan Reh 2004).

Silica-Induced Respiratory Diseases

Planning and Production Inputs

RDRP research on silicosis has addressed six primary objectives:

- Reduce deaths due to silicosis through research and policy development.
- Promote use of silica substitutes for abrasive blasting.
- Assess silica exposures and controls.
- Provide relevant disease progression and dose-response data for standard-setting groups.

- Demonstrate that freshly fractured silica is associated with a highly reactive dust.
- Demonstrate that oxidant injury is a critical mechanism for interstitial lung disease.

As late as the mid-1980s, NIOSH estimated that 2.3 to 4.3 million workers in a wide variety of mining and nonmining occupations were still at risk for occupational exposure from silica dust. By 1991, the RDRP had established a system for national surveillance both to document silica exposures and to estimate the prevalence of silicosis. According to more recent NIOSH data (2006d), between 1968 and 2003 there were 15,714 certified deaths that mentioned silicosis and annually reported deaths due to silicosis decreased from 1,065 in 1969 to 179 in 2003. Mortality from silicosis has declined, with the total number of silicosis deaths from 1999 to 2003 at 52% of the corresponding number for the period 1985-1989 (NIOSH 2006d). Nonetheless, the problem persists. It has been estimated that between 3,600 and 7,300 new cases of silicosis are recognized each year (Rosenman et al. 2003). Moreover, OSHA and MSHA silica sampling data (1999) showed that 47% of samples from construction, 38% from manufacturing, and 8% from mining exceed their PELs for silica (NIOSH 2006a). More than half the silica samples in construction and manufacturing exceed the NIOSH recommended exposure limit (REL) for silica (0.05 mg/m^3) (NIOSH 2006a).

Activities, Outputs, and Outcomes

RDRP scientists have documented increased toxicity from freshly fractured silica, which may elevate the risk of silicosis among miners, rock drillers, sandblasters, and silica flour millers (Castranova et al. 1996a). They have also produced hundreds of abstracts, presentations, and publications on silica (249 published in the last 10 years) (NIOSH 2006a). On the basis of RDRP research, NIOSH has published 12 official policy documents on silica in the workplace that began with the 1974 criteria document "Occupational Exposure to Crystalline Silica" completed and disseminated in 1974 (NIOSH 1974). This document set an REL for crystalline silica of 0.05 mg/m^3. In 2002, NIOSH reassessed its silica data, updated its risk assessment, and reaffirmed the NIOSH REL of 0.05 mg/m^3 for silica (half that of the OSHA PEL of 0.1 mg/m^3) (NIOSH 2002a). NIOSH cosponsored International Symposia on Silica, Silicosis and Cancer in 1993 and 2002 and has used interagency agreements to conduct several research projects designed to address essential knowledge gaps in support of a new silica standard (Perry 2006). NIOSH (2006a, p. 69) states that "OSHA intends to ask RDRP scientists for an expert review of OSHA's anticipated new silica standard prior to its release for public comment."

In response to documented high rates of silicosis and acute silicosis arising from sandblasting, NIOSH published an alert "Request for Assistance Preventing Silicosis and Deaths from Sandblasting" (NIOSH 1992a). The alert recommended use of abrasive substitutes for sand, but the economic feasibility of substitutes and the potential for toxicity of substitute abrasives were not understood fully at the time. Substitute materials for silica sand and abrasive blasting were evaluated through controlled test blasting. The cost of substitute materials was evaluated and the effectiveness of substitute materials was determined. RDRP results documented that pulmonary toxicity of abrasive substitutes (specular hematite and steel grit) was less than that of sand (Craighead et al. 1988; Hubbs et al. 2001, 2005; Porter et al. 2002). These findings were documented in several peer-reviewed reports and three U.S. Department of Commerce National Technical Information Service publications and were communicated to industrial hygiene professionals and government regulatory agencies (KTA-Tator 1998). As a result of RDRP research, use of abrasive substitutes for sand increased, while the use of silica sand for sandblasting decreased by 47% (between 1996 and 2004). OSHA considers information on the toxicity of sandblasting substitutes as critical to its ongoing effort to develop and promulgate a new silica standard (NIOSH 2006a). Consistent with NIOSH's efforts, the National Toxicology Program (NTP) included selected abrasive blasting substitutes for a subchronic inhalation study (2002).

The RDRP partnered with the American Lung Association in a joint campaign on silicosis prevention that began in 1996. This campaign engaged multiple partners from industry, labor, and academia throughout the United States and in several foreign countries. The RDRP developed engineering controls and recommendations to reduce occupational exposures to silica that resulted in 80 publications on the control of silica exposures in several industry sectors and mining. On the basis of the RDRP joint campaign on prevention of silicosis, OSHA developed a Special Emphasis Program to reduce silicosis. Similarly, MSHA undertook efforts to reduce occupational exposures based on RDRP research that included the Miners Choice Examination Program for surface coal miners who are at particular risk for silicosis associated with drilling overburden rock. A World Health Organization (WHO) report authored by the RDRP was instrumental in launching the ongoing ILO/WHO International Programme on Global Elimination of Silicosis (1995).

In 1998, OSHA asked NIOSH to collect data on silica exposure and engineering controls in a variety of workplaces to determine the technical feasibility of a proposed rule. These environmental studies, in both construction and general industry, by RDRP scientists have documented significant exceedances of the NIOSH PEL in both construction (36% of samples) and general industry (16% of samples). Therefore, RDRP studies focused on evaluating engineering controls in a variety of construction and general industry settings. Results have documented prevention measures for silica exposures that led to peer-reviewed papers documenting the technological feasibility, cost, and impact of these measures (Echt et al. 2002, 2003; Echt and Sieber 2002). RDRP research continues to evaluate engineering control technology for selected silica-generating activities, development of a silica-focused workplace solutions document to provide practical guidance for dust control, and transfer of engineering control knowledge from mining to selected general industry applications (Flanagan et al. 2003; Flynn and Susi 2003; Croteau et al. 2004).

Data used to set standards are based on available population-based studies of silicosis in working populations. The development of biomarkers may provide evidence of an even earlier stage of disease that could benefit prevention efforts. Toward resolution of these issues, RDRP scientists have undertaken a major inhalation study of silica-induced pulmonary responses in a rat model. This multidisciplinary study is intended to identify a threshold lung burden beyond which disease will progress without further exposure (Green et al. 1989; Kuempel et al. 2001). Studies have also been undertaken in humans and have led to the identification of genetic polymorphisms associated with an increased incidence of silicosis (Yucesoy et al. 2001). Identification of biomarkers that can indicate potential susceptibility to silicosis or provide early evidence of the disease is suggested as a useful adjunct to the diagnosis of silicosis. In addition to continuing consultation with OSHA, research on biomarkers continues through a collaborative study with South Africa's National Center of Occupational Health and also through a collaborative study with the People's Republic of China (silicosis in tin and pottery workers) with attention to particle surface characteristics in these industries. Finally, RDRP scientists are investigating dose- and time-dependent silica-induced lung cancer in a susceptible mouse model and are completing a study of the role of a model gene (mineral dust-induced gene, *mdig*) in silica-induced lung cancer to provide mechanistic data on silica-induced lung cancer.

It is known that certain occupations, including sandblasting, rock drilling, silica flour milling, concrete cutting, and tunneling are associated with a high incidence of silicosis. A fundamental question addressed by RDRP activities is whether this high rate is due to high concentrations of airborne dust or whether freshly fractured silica particles are inherently more toxic than aged silica particles. Certainly, this mechanism has long been suspected to be relevant to disease causation, but the question remains unsettled. In vitro exposure of lung cells to freshly fractured versus aged silica particles has revealed freshly fractured silica to be more cytotoxic (Vallyathan et al. 1995; Castranova et al. 1996b). Coating these particles with an organosaline material effectively coats freshly fractured silica and mitigates cytotoxicity and thereby represents a potential prevention strategy in certain occupational exposure settings. RDRP scientists also have worked closely with MSHA to develop engineering controls in underground mines and to use dust capture technology on drills.

These control technologies have been widely disseminated in the form of technical reports and peer-reviewed publications relevant to sandblasters, rock drillers, construction workers, and miners (NIOSH 1992a,b; Gulumian et al. 2006).

A critical mechanistic issue addressed by the RDRP is the hypothesis that the pathogenic potency of a particle depends in part on its ability to generate reactive species on its surface or to stimulate production of reactive oxygen and nitrogen species from phagocytic cells. RDRP scientists have developed techniques to measure the production of reactive oxygen species and reactive nitrogen species and have also developed techniques to monitor oxidant stress. This research has been aided by experimental in vivo exposure studies in the rat. This line of research has also demonstrated that silica-induced oxidants can activate signaling pathways for the production of inflammatory cytokines and chemokines, fibrosis, and growth factors. It is hoped that elucidation of the mechanistic role of oxidative species will result in identification of biomarkers of silicosis and silica-induced cancer. Four important peer-reviewed publications and multiple abstracts and presentations have resulted from this research (Shi et al. 1998; Castranova 2004; Zeidler and Castranova 2004; Porter et al. 2006). In addition, RDRP scientists sponsored or co-sponsored three conferences on oxidant injury in 1993, 1997, and 2002. RDRP scientists have also cowritten four books on oxidant injury (Van Dyke and Castranova 1987; Castranova et al. 1996a; Vallyathan et al. 2002, 2004). These findings have helped the scientific community and regulatory agencies to better understand the mechanism for initiating the progression of silicosis. Oxidant stress has also been identified as an important paradigm for nanoparticle toxicity (Nel et al. 2006).

RDRP expertise has been useful to external bodies, including the International Agency for Research on Cancer (IARC) monograph, "Evaluation of the Carcinogenic Risk of Chemicals to Humans. Silica, Silicate, Coal Dust, and Para-aramid Fibers" (IARC 1997), and to the EPA and the International Life Sciences Institute in their effort to develop short-term screening strategies for fiber toxicities and nanomaterials.

RDRP research will continue to focus on the role of oxidant stress and occupational lung disease with an emphasis on developing inhibitors, therapeutic interventions, and delivery systems to treat pulmonary disease. Current program efforts for silica-induced respiratory disease research are (1) to provide necessary research to inform OSHA in anticipation of its new silica standard; (2) to reduce silica exposures through increased substitution, development, and dissemination of guidance for dust control in at-risk industries and occupations; (3) to continue silica and silicosis surveillance; (4) to provide mechanistic data on dose- and time-dependent silica-induced lung cancer; (5) to provide mechanistic data on the role of a novel gene (*mdig*) in silica-induced lung cancer; and (6) to develop tools for early detection of silica-induced respiratory disease, including the use of digital radiography in silicosis surveillance and development of a possible biomarker to help detect early development of disease.

In addition, although specifically relevant to RDRP work on cancer risk generally, it should be noted here that NIOSH research supported a link between silica exposure and lung cancer through numerous epidemiologic studies, HHEs, and laboratory investigations that focused on silica exposure in the 1970s and 1980s. IARC cited this research to support its determination that crystalline silica, in the form of quartz and cristobalite, were probable human carcinogens (IARC 1997). Further, interagency agreements with OSHA in 1999, 2000, and 2002 have funded RDRP projects on silicosis in an animal model to evaluate responses at the lower end of the dose-response scale and to evaluate the ability of silica to cause lung cancer in a susceptible animal model.

Assessment of Relevance

The wide range of occupations and activities with silica exposures, the severe health effects of silicosis, and a direct causal link to occupational exposures all underscore the high relevance of this focus area to NIOSH's core mission and the RDRP's specific goals. In that context, the goals of the RDRP for silica-induced lung diseases focus on key areas that can contribute to a better understanding of the adverse health effects of silica and to better protect workers. These goals include laboratory research on a mechanistic understanding of silica's toxicity and applied field research relevant to silicosis surveillance and monitoring occupational exposures. Both foci should assist in the effort to develop and promulgate a revised silica standard as part of a broader effort to reduce the incidence of and mortality from the disease. Unfortunately, with the exception of cancer risk assessment, there does not appear to be a complementary epidemiologic effort targeting silicosis and other silica-associated health that is integrated with the laboratory and applied field activities delineated above.

Assessment of Impact

RDRP work to achieve goals related to reducing silicosis and other silica-induced respiratory diseases has had a measurable impact. As described above, NIOSH reports that, while annual silicosis deaths decreased from 1,065 in 1968 to 179 in 2003, recent estimates show that between 3,600 and 7,300 cases of silicosis are newly recognized each year. Mechanistic, dose-response, and epidemiologic research conducted or sponsored by NIOSH have enhanced the understanding of silica-related disease processes in support of the goal of reducing these diseases. The mechanisms driving disease causation have been better defined through a series of research efforts on silica-related oxidant stress. Research on freshly fractured silica has demonstrated its enhanced toxicity, mechanisms driving that effect, and means to reduce those toxic effects.

RDRP-applied research efforts have helped to reduce the use of silica in sandblasting operations and have driven the development of alternatives for silica sand in these operations and, as described above, NIOSH (2006a) indicates that the use of silica sand for sandblasting decreased by 47% between 1996 and 2004. Nonetheless, even a 50% reduction in use means that only half the problem has been eliminated. This is particularly noteworthy since, in the same time frame, silica-based abrasive blasting has been virtually eliminated within the European Union. Furthermore, NIOSH analysis of exposure monitoring for a variety of industries has indicated that levels of exposure frequently exceed even current standards, suggesting that any new standard, absent changes in work practices and enforcement, is unlikely to achieve adequate control. Overall, deaths attributable to silica have declined substantially since 1983, but, once again, it is notable that this wholly preventable cause of mortality continues to occur so frequently as appears to be documented by NIOSH.

Fiber-Induced Interstitial Lung Disease

Planning and Production Inputs

NIOSH's very first criteria document was "Criteria for a Recommended Standard: Occupational Exposure to Asbestos" (NIOSH 1972). This criteria document was instrumental in moving OSHA and MSHA to establish occupational PELs for asbestos. Despite regulations and because of the thousands of products and processes that use asbestos, even as late as the 1980s, NIOSH estimated that 1.5 million workers remained at-risk for asbestosis (NIOSH 1986b). National mortality data documented that asbestosis mortality increased in the United States from 100 deaths in 1968 to nearly 1,500 in 2002, likely, in part, to a long latency and also to increased recognition and diagnosis of the disease. OSHA and MSHA documented substantial declines in exposure for manufacturing, construction, and mining over the past two decades (NIOSH 2003). After tackling asbestos early on, NIOSH later shifted focus to emerging fiber-induced lung diseases from RCFs and carbon nanotubes, asbestos-contaminated vermiculite, and respirable nylon flock fibers. Long-standing NIOSH research strategies include the "NIOSH Interdivisional Fiber Subcommittee Final Report" (Baron et al. 1999):

- Validate animal models to provide a quantitative relationship between animal and human responses to fiber exposures.
- Characterize and quantify fiber exposure in the workplace.

- Design size-selective technology to produce homogeneous fiber samples for study.
- Determine mechanisms of action, rates of dissolution, and dose-response relationships for various homogeneous fiber samples.

Although a great deal is known about the adverse health effects of inhaled fibers, much of it with asbestos as a prototype hazard, new exposures continue to emerge that may put workers at risk. At issue is the development of a universal definition for fibers and standard criteria for measuring them. Methods for producing fiber samples of uniform dimensions are an intermediate objective as it would facilitate experimental toxicology research.

Activities, Outputs, and Outcomes

Vermiculite

The vermiculite episode serves as a good illustration of a key set of NIOSH activities and outcomes in relation to fibers. A now closed mine near Libby, Montana, accounted for three-fourths of the worldwide vermiculite production for many decades during the last century. Vermiculite, once expanded, was used as loose-fill thermal insulation in residential and commercial buildings and was also used in horticulture and many other applications. This vermiculate ore was later documented to be substantially contaminated with actinolite-tremolite asbestos and other closely related asbestiform amphibole fibers. RDRP engagement in this research began as far back as 1979 (NIOSH 2006a). Since that time, RDRP staff have worked with the EPA and the Agency for Toxic Substances and Disease Registry to communicate and to assess potential community health risks from asbestos-contaminated vermiculite. At OSHA's request, RDRP investigators undertook field studies to determine current levels of occupational exposure to vermiculite through asbestos fibers in expansion plants for agricultural uses. More recently, RDRP scientists used findings from their studies of asbestos-contaminated vermiculite to document and support development of a MSHA PEL for asbestos applicable to the mining industry to bring it in line with the more protective OSHA PEL for asbestos in general industry.

RDRP staff continue to support MSHA in the current effort to establish a more protective "Proposed Rule on Asbestos Exposure Limit" (NIOSH 2005a). The RDRP is now engaged in updating and expanding its 1980s cohort mortality study of vermiculite workers to examine a broader range of exposures, including short-term workers with low cumulative exposures to fiber, to examine disease outcomes not previously documented, and to provide more precise risk estimates for low-level exposures to fiber. This research has implications for both occupational and selected community exposures.

Nylon Flock

In 1996, an employer used the HHE/TA to request that the RDRP investigate multiple cases of biopsy-documented interstitial lung disease among workers at a small Rhode Island textile plant that produced nylon flock and flock-coated upholstery fabric (Washko et al. 1998). RDRP investigators conducted a comprehensive environmental sampling and medical survey, and RDRP laboratory-based investigators analyzed the airborne dust and conducted animal studies to characterize its respiratory toxicity. Brown University and RDRP scientists reported a new interstitial lung disease in MMWR, a disease dubbed "flock worker's lung" (MMWR 1997). RDRP investigators later conducted additional HHEs at other U.S. flock plants and documented that subclinical disease among these workers was both widespread and associated with time worked per week, specifically in flocking processes that used compressed air between production runs (Daroowalla et al. 1998a,b; Antao and Piacitelli 2004; Piacitelli and Antao 2006). RDRP laboratory research documented that respirable fibers were arising from cutting nylon tow filaments when rotary cutters were not optimally sharpened and aligned, which resulted in melting and tailing nylon ends of cut flock becoming airborne with subsequent milling (Burkhart et al. 1999). Animal models documented an intense inflammatory response in their lungs (Porter et al. 1999). Other RDRP investigators recommended prevention measures including improved maintenance, exhaust ventilation, limited use of compressed air, and a respiratory surveillance program with ongoing medical screening.

After the initial MMWR article, RDRP investigators completed additional HHEs. They also published several peer-reviewed papers pertaining to flock exposures and interstitial lung disease (Burkhart et al. 1999; Porter et al. 1999; Washko et al. 2000; Daroowalla et al. 2005) and hosted a workshop on "Flock Worker's Lung" that was summarized in two published papers (Boag et al. 1999; Eschenbacher et al. 1999). RDRP investigators also hosted several meetings with representatives of the major domestic producers of nylon tow to inform them of this newly recognized hazard and discussed product stewardship. In addition, they made several presentations to annual meetings of the American Flock Association and at an international trade association meeting on flock. Follow-up by RDRP investigators at the Rhode Island textile mill found no new cases of flock worker's lung after exposure controls were implemented.

Refractory Ceramic Fibers

Manufactured RCFs are increasingly used as substitutes for asbestos in many industrial products and applications. RCFs have been shown to cause mesothelioma and lung cancer in animal models and have been associated with pleural plaques and reduced lung function among exposed workers in a cross-sectional study (Lemasters et al. 1998). An estimated 31,000 workers in the United States are exposed to RCFs. RDRP investigators have studied airborne RCFs since the early 1980s but more recently developed and disseminated a NIOSH criteria document "Criteria for Recommended Standard: Occupational Exposure to Refractory Ceramic Fibers" (NIOSH 2006e) and have communicated their findings to industry, academia, organized labor, and other federal agencies. This document recommends a comprehensive worker health protection program and an exposure guideline of 0.5 fiber/cm^3. Further, NIOSH has supported both the EPA and OSHA in monitoring exposure and the RCF industries' efforts to reduce exposures. RDRP scientists have documented the efficacy of engineering controls and have transferred their findings to fiber-manufacturing industries, labor interests, academia, and OSHA (Dunn et al. 2004). RDRP staff contributed to the development of an industry product stewardship program that provided guidelines for exposure monitoring, a medical surveillance program, and a recommended exposure guideline for RCFs. The RCF coalition has reportedly met or exceeded all product stewardship plan requirements and annually reports its progress to RDRP scientists. The RDRP plans to continue to monitor progress of the stewardship plan and participate in annual briefings. As a result, RDRP scientists have documented a dramatic decrease in exposure (NIOSH 2006a) and improved lung function among workers who manufacture RCFs in a longitudinal study (Lockey et al. 1998).

Mechanisms of Fiber Toxicity

RDRP scientists have continued to study asbestos mechanisms from both a mechanistic and an applied field research perspective. For example, they have published scientific papers on the role of oxygen radicals in asbestos toxicity on a dielectrophoresis method for sorting fibers by length and on the relationship between fiber toxicity and fiber length (Simeonova et al. 1997; Blake et al. 1998; Deye et al. 1999). RDRP staff have also participated in numerous panels and workshops and have provided technical advice to the EPA, the Agency for Toxic Substances and Disease Registry, and WHO. The EPA has adopted long-term and short-term fiber-testing strategies to which RDRP scientists have contributed, and a new EPA standard is being formulated. RDRP scientific outputs have also influenced IARC ranking of fibers as to carcinogenicity in humans. Current initiatives include RDRP efforts to reanalyze an unspecified "cohort mortality study of asbestos textile workers originally published decades ago" with characterization of exposures to include additional information on fiber dimensions using stored air sample filters (NIOSH 2006a). RDRP scientists are also engaged in drafting a white paper on nonasbestiform mineral "cleavage fragments." This white paper is intended to drive policy and to recommend research to identify gaps in knowledge.

Assessment of Relevance

Fiber-induced diseases have long been a serious heath risk in occupational settings. Research on occupational exposures resulting in fiber-induced disease and the underlying biologic mechanisms is unquestionably of high relevance to occupational safely and health. Fibrotic lung disease, pleural disease, and thoracic malignancies are strongly related to exposure to asbestos fibers, and, unfortunately, it is clear that this is not the only occupational fiber with potential risk of lung fibrosis. The RDRP's research on occupational exposures to asbestiform fibers at a Libby, Montana, vermiculite mine (and industries using the ore) proved immensely important in stemming exposures and disease related to asbestos-contaminated vermiculite. The RDRP's evaluation of lung disease in textile workers exposed to nylon fibers in dust from nylon flock is also an important study area because of unknown disease etiology, occurrence, and applicability of the findings to other sites. Efforts to document lung disease from exposure to other synthetic fibers are also highly relevant RDRP research on natural and synthetic fibers likely to elicit toxicity; to identify the biologic mechanisms driving that toxicity, it is necessary to better understand disease causation and develop protective measures.

Assessment of Impacts

The RDRP's efforts related to fiber-induced diseases have had a substantial impact on worker safety and health, with decreases in exposures to causative agents, decreases in disease incidence, important findings about the etiology and biology of these lung diseases, and improvements in worker health and safety. Research on a cluster of workers with pleural disease contributed to the finding that vermiculite ore from a Libby, Montana, mine was contaminated with asbestos. The mine was later closed based partly on these findings and on RDRP research on exposures and epidemiology at the mine site. RDRP research also resulted in the determination that exposure to nylon fibers was associated with lung disease in nylon flock workers. Beyond the original site that was investigated, additional cases of "flock worker's lung" were identified. Nylon flock dust exposures are now recognized as a preventable health hazard. NIOSH reports that the lack of new cases of "flock worker's lung" from the company initially investigated and among U.S. flock workers suggests that incidence of the disease has been reduced (NIOSH 2006a), although additional surveillance is clearly warranted. RDRP projects on engineering controls to reduce exposure to flock dust have also helped reduce RCF exposures to workers and show promise for further reducing the incidence of disease. An RCF product stewardship program contains exposure monitoring, medical surveillance, and exposure guidelines. NIOSH states that RCF fiber concentrations in manufacturing settings have declined dramatically and that these reduced exposures have been reflected in improved pulmonary function tests by workers (NIOSH 2006a). Here too, further follow-up is particularly relevant.

Chronic Beryllium Disease

Planning and Production Inputs

It is known that compliance with the current OSHA PEL does not ensure prevention of beryllium sensitization or progression to CBD. The relationship between exposure and disease, still not fully understood, suggests that other characteristics of beryllium exposure (particle size, chemical form, and the contributions of dermal contact) are likely relevant. It is also recognized that field portable detection methods for beryllium are needed for fast, on-site analysis.

In an effort to study the determinants of both beryllium sensitization and progression to CBD among exposed workers, NIOSH has supported multiple extramural epidemiologic studies (Newman et al. 2005; Rosenman et al. 2005). From an intramural perspective, RDRP scientists have collaborated with Brush Wellman Inc., the sole manufacturer of beryllium in the United States, to conduct epidemiologic studies, collect airborne and other work process samples, and assess transfer of their research to practice. This program has three major initiatives: medical surveillance and epidemiology, exposure assessment and bioavailability, and genetic studies.

Activities, Outputs, and Outcomes

Since 1996, the beryllium disease program has produced 54 peer-reviewed journal articles and has presented its findings through several national and international meetings. Epidemiology and medical surveillance research has documented that copper-beryllium alloys make up the most widely used form of beryllium. Studies of workers with exposure to copper-beryllium alloy show low levels of sensitization and CBD together with relatively low concentrations of airborne beryllium (Schuler et al. 2005). RDRP scientists have also documented that the comprehensive prevention program implemented by Brush Wellman in 2000 has successfully reduced, albeit not eliminated, the incidence of beryllium sensitization compared with the preintervention sensitization rate (Cummings et al. 2007). These findings suggest that a comprehensive approach to beryllium control and prevention, including dermal protection, is essential for effective worker protection.

RDRP research on physicochemical characterization of airborne beryllium has been widely recognized for its characterization of particle behavior (Stefaniak et al. 2003, 2004). RDRP scientists have also developed a highly sensitive, field portable analytical method for determining trace beryllium in workplace air and in surface samples (Agrawal et al. 2006).

RDRP research has provided a new understanding of the underlying basis for immunogenetic observations of CBD risk (Snyder et al. 2003; Weston et al. 2005). This body of research has uncovered a hierarchy of risk among exposed workers with the single genetic polymorphism most frequently associated with CBD. To better understand this hierarchy of risk, RDRP scientists have collaborated with investigators at Lawrence Berkeley National Laboratory to develop transgenic mouse models with different degrees of risk for beryllium sensitization and granulomatous inflammation that will allow experimental gene-exposure and gene-gene interaction studies to be conducted.

RDRP researchers have forged an innovative partnership with Brush Wellman that could serve as a model for cooperative NIOSH-industry research activities in other sectors. RDRP investigators communicate their findings in regular meetings with Brush Wellman but also via peer-reviewed publications and technical reports to other interested parties including unions, other industries dealing with similar workplace issues, government agencies, and university researchers. RDRP staff have also worked with Brush Wellman to develop computer-based decision-making software to help assess exposure factors and workplace-specific decisions.

According to the evidence package (NIOSH 2006a), RDRP is producing a NIOSH Alert "Preventing Beryllium Sensitization and Chronic Beryllium Disease" that will be mailed to all known workplaces that use beryllium and will be posted on the NIOSH website. RDRP scientists have cosponsored several beryllium conferences, including a Department of Energy Beryllium Research Symposium, a National Jewish Research Center Conference, and an international beryllium research conference in Montreal. OSHA is in the process of developing a standard for beryllium for which OSHA analysts have access to NIOSH epidemiologic data.

In 2006, RDRP scientists submitted a program proposal for funding through fiscal year 2010 that includes four projects: (1) continued epidemiologic analysis of the Brush Wellman Preventive Program to extend the assessment of the risk of sensitization at other facilities, the risk of CBD at low levels of sensitization, and the relationship between genetic markers and disease risk; (2) completion of physicochemical characterization of beryllium materials with the goal of determining the bioavailability of beryllium to the body and development of a better approach to assess the risk of sensitization and CBD risk; (3) continued analysis of alternative exposure pathways, specifically the dermal route of exposure and its relationship to sensitization risk; and (4) a set of experimental studies with a transgenic mouse model (NIOSH 2006a).

Assessment of Relevance

This is a highly focused objective. However, in the broader sense, understanding sensitization to beryllium, as well as the progression to CBD, is highly relevant in occupational safety and health for a number of reasons. Significant numbers of workers are exposed to beryllium, leading to a severe, potentially lethal disease. In addition, CBD can serve as a model for studying granulomatous lung disease in general. This is relevant to other workplace metals as well as to sarcoidosis, an otherwise idiopathic lung disease. The RDRP beryllium research program includes the necessary components—medical surveillance and epidemiology, exposure assessment and bioavailability, and genetic markers—to better understand disease causation and to protect worker health.

Assessment of Impact

The RDRP's research has been of great importance in understanding the development of CBD and in protecting workers. The work has resulted in a better understanding of the complexities of beryllium's dose-response on exposed individuals. The RDRP has developed analytical technologies for determining trace beryllium concentrations in the workplace, has focused on genetic markers responsible for increased risk and sensitization to beryllium, and has collaborated in the development of transgenic mouse models with different susceptibility to sensitization. The program has contributed substantially to the peer-reviewed literature on multiple aspects of beryllium exposure and disease. An innovative cooperative research program with the country's major beryllium producer, which includes applying findings in that producer's facilities, has showed promising results in decreasing beryllium sensitization to workers; the improvement presumably results from engineering controls and use of protective equipment to control exposures. An important but major challenge will be to extend this model to exposure to other metals and, potentially, to unexplained occupational clusters of sarcoidosis.

STRATEGIC GOAL 3: PREVENT AND REDUCE WORK-RELATED INFECTIOUS RESPIRATORY DISEASES

Introduction

This section reviews the Infectious Diseases Program (IDP) of the NIOSH RDRP. To facilitate the review, we treat the tuberculosis (TB) program and the engineering controls/respirator technology as separate entities, since the TB program has a long history of accomplishment and engineering controls/respirator technology affects all RDPD programs. All other subprograms are considered as a group, since they are newer (anthrax, emerging infections, understanding the effect of occupational exposures on pulmonary susceptibility to infection) or are of historical interest only (histoplasmosis). The RDRP strategic goal is to prevent and reduce occupationally related respiratory infectious diseases. To achieve this strategic goal, the following intermediate goals have been developed (NIOSH 2006a):

- Maintain reductions in occupational incidence of TB in high-risk work settings.
- Protect workers from bioterrorism agents.
- Protect workers from occupational acquisition of emerging diseases (including SARS, avian flu, and pandemic flu).
- Protect workers from occupational exposures that make them susceptible to respiratory infections.
- Prevent outbreaks of occupational histoplasmosis by maintaining worker and employer awareness.

Table 2-4 summarizes the outputs, intermediate outcomes, and end outcomes for each of these intermediate goals.

The various subgoals relating to specific diseases (TB, bioterrorism agents, SARS, avian and pandemic influenza, and histoplasmosis[2]) under this strategic goal do not provide an overarching program structure for preventing emerging infectious diseases within work environments. However, several themes would provide such structure. For example, work on engineering controls and personal respirator technology, the current subgoal to understand mechanisms of pulmonary susceptibility to infection resulting from occupational exposures, and the need for surveillance to detect work-related infectious respiratory diseases extend across the disease-specific subgoals.

[2]It is unclear why histoplasmosis is included as part of the strategic goal. No further research is planned, nor are there specific surveillance programs targeted toward identifying outbreaks.

TABLE 2-4 Summary of IDP Outputs and Outcomes[a]

Area	Outputs and Transfer
TB	– 135 reports, abstracts, chapters, documents, videos, and other reports published and disseminated. – International training activities conducted in collaboration with other parts of CDC included environmental controls. – Protect yourself against TB, a respiratory protection guide for health care workers (94,935 copies distributed). – TB respiratory protection program in health care facilities, administrator's guide (12,839 video, 55,413 print copies distributed). – TB prevention and control in correctional and detention facilities (2006). – >100 HHEs since 1990.
Anthrax	– 46 journal articles, chapters, HHE reports published. – Recommendations for personal protective equipment (PPE) and engineering controls developed and disseminated to stakeholders. – Developed sampling methods including *B. anthracis* culture. – Conducted remediation (USPS and Capitol Hill), provided technical support for decontamination workers, and trained FBI and Coast Guard staff. – Validated a new biological detection system (BDS) and a new ventilation-filtration system (VFS) for USPS facilities. – Contributed to development of many educational products disseminated by the CDC. – Developed two major guidance documents on building protection against chemical, biological, radiological, and nuclear (CBRN) attacks. – Developed standards for CBRN certification of respirators.
SARS, Avian and Pandemic Influenza	– Through direct involvement on CDC's SARS teams: • Responded to requests for help from Canada and Taiwan. • Screened incoming passengers at domestic airports. • Developed and disseminated information, including 8 documents and 7 websites. – Through direct involvement with other groups in the CDC and beyond (e.g., OSHA), helped to develop a range of recommendations that have been disseminated in the form of 5 documents and 9 websites.
Occupational exposures and susceptibility to infections	– 33 peer-reviewed publications showing that diesel exhaust, welding fumes, and silica exposures increase susceptibility of rodents to experimental respiratory infections. – Welding fume generator with robotic welding arm developed to produce controlled exposures for animal studies. – DoD entered into inter-agency agreement to determine effect of mixed dusts on asthma and lung infection.
Histoplasmosis	– 2 HHE reports, 2 peer-reviewed reports, an MMWR report, a trade journal report, an English/Spanish fact sheet, and a NIOSH numbered publication ("Histoplasmosis: Protecting Workers at Risk") published.

[a]Most of the data in this table were supplied by NIOSH in response to a specific request from the committee. Data were supplemented from other materials, as appropriate.

Intermediate Outcomes	End Outcomes and External Factors
– CDC incorporated recommendations on environmental controls into guidance on TB prevention for the health care and correctional and detention settings. – Influenced guidance from the American Institute of Architects and the American Society of Heating, Refrigerating, and Air-conditioning Engineers. – Influenced standards used by the Joint Commission on Accreditation of Healthcare Organizations and by OSHA.	– Overall number of TB cases in U.S. declined by 46% between 1992 and 2004. – Incidence of active TB among health care workers declined from 5.6 to 4.6 per 100,000 (1994-1998). – No good data for assessing occupationally related TB in other worker groups and settings.
– The CDC and Health and Human Services rapidly adopted and posted preventive recommendations in the immediate aftermath of the "anthrax attacks." – USPS facilities were decontaminated following recommendations and the recommended VFSs and BDSs were installed promptly.	– While it is impossible to be certain, it is likely that the death toll from the "anthrax attacks" would have been more than 5 had comprehensive preventive guidance not been developed. – Though not measurable, preventive guidance is also intended to contribute to safer workplaces in the event of future biological attacks.
– OSHA and others used CDC guidance in developing their recommendations. – Avian influenza guidelines implemented by poultry producers. WHO revised recommendations to include use of designated disposable respirators that provide protection equivalent to NIOSH-certified N95 respirators.	– Only 8 people in the U.S. contracted SARS. – If prevention guidelines are followed, some impact on reducing morbidity and mortality of avian-pandemic flu can be expected if it strikes the U.S. population. – Revised WHO respirator recommendations reduce the potential risk of overrunning the supply of NIOSH-certified N95 respirators should pandemic influenza occur.
– The EPA produced "Health Assessment Document for Diesel Engine Exhaust." – NTP plans to use welding fume generator system for chronic exposure studies.	– Findings of research on increased susceptibility to lung infection can be expected to lead to improved protections for workers, with resulting reductions in lung disease among workers at risk.
– Recommendations adopted as guidance by OSHA, Center to Protect Workers' Rights, various state health departments, and company occupational safety and health programs.	– Occupational outbreaks of histoplasmosis are infrequent. – Established prevention recommendations remain relevant.

Tuberculosis

Although not stated in formal terms, the goal of the TB program is to "prevent and reduce occupationally related TB within the context of the broader CDC [Centers for Disease Control and Prevention] response to eliminate TB in the U.S." and follows the strategic plan for TB set out, updated, and implemented by the CDC (MMWR 1989a, 1992). In addition to other CDC guidance documents (CDC 2002; MMWR 2005, 2006b), a 2001 report of the Institute of Medicine has served to provide program objectives (IOM 2001). Taken together, these sources provide an extensive body of guidance for the NIOSH TB program. To a large extent, these strategic goals appear to be driven largely, and appropriately, toward national priorities that were established through the CDC.

The NIOSH TB program has brought specific expertise to the occupational setting, especially in areas of respirators and environmental controls. Lack of specific occupational TB surveillance programs represents a major challenge to the RDRP goal. Moreover, many of the surveillance systems on which the TB program relies (e.g., Bureau of Labor Statistics Surveys) do not provide detailed information about specific occupations in their electronic databases. Given the continued immigration of documented and undocumented laborers from areas of TB endemicity, more specific occupational TB surveillance could help better target NIOSH's resources. The lack of such information remains a critical unmet need.

The NIOSH FY2007 project planning guidance document does not list any specific program related to TB. In the evidence package, respirator research is listed as a major priority area. This narrow focus carries over into the TB program's role in the IDP environmental microbiology program. In addition, there is no other mention that future research plans are emerging in areas that are unique to the TB program.

Planning and Production Inputs

The TB program has functioned in the context of the CDC's overall program for surveillance and prevention of the spread of TB, as noted in the preceding paragraph. In addition, appendix material that accompanied the evidence package documents 22 HHEs from 1994 to 2004 that were completed in response to stakeholder requests related to TB. Over the period of evaluation, the TB program has responded to more than 100 requests for help in TB control, which include 43 HHEs. Requests have come from health care facilities, homeless shelters, and correctional and social service facilities.

Activities, Outputs, and Outcomes

The program lists a large number of outputs and transfers in the form of publications, abstracts, book chapters, videos, and a variety of other reports (135 in all). Of particular note were contributions to the development of environmental controls with respect to respiratory technology and use of an upper-room ultraviolet germicidal irradiation (UVGI) system to inactivate or kill airborne TB bacteria (and molds/spores) (Miller et al. 2002; Xu et al. 2003, 2005) and work on respiratory protection focused on surgical masks (Qian et al. 1997, 1998; Johnson et al. 1998) and fitting characteristics of face pieces (Coffey et al. 2004, 2006). Publications related to exposure assessment have reported the development of rapid detection methods for *Mycobacterium tuberculosis* and methods to document concentration and size distribution of infectious particles.

Although the IDP has identified health care, correctional, and migrant workers as occupational groups of special interest with respect to occupational TB, the IDP does not supervise its surveillance activities with respect to these groups. The general issue of surveillance is addressed in the context of the overall evaluation of the IDP (see below). However, the IDP has worked with health departments to address these issues in particular cases. For example, NIOSH SENSOR (NIOSH 1997b) supported the California Department of Health Services in a program of targeted TB surveillance among prison and other correctional employees during the resurgence of TB associated with the human immunodeficiency virus in the mid-1990s. Nonetheless, systematic TB surveillance in high-risk occupational settings depends on other programs for data.

Intermediate outcomes are listed in Table 2-4. The American Institute of Architects adopted and updated CDC "Guidelines for Design and Construction of Hospitals and Health Care Facilities" to which the IDP provided expertise on isolation rooms to prevent the spread of infections in these facilities (AIA/FGI 2006). CDC guidance documents for preventing the spread of TB have been prepared with input from the IDP (MMWR 2005, 2006b) and, with respect to specific IDP inputs, these documents contain information on respirator protective gear and ventilation systems in health care facilities.

In terms of end outcomes, much of the IDP TB control efforts exist within the context of broader efforts by the CDC and other agencies. Therefore, it is not possible to provide a quantitative estimate for what percentage of the decline in TB, in general and in occupational settings, can be attributed to the IDP. However, it is reasonable to presume that these programs depend on the unique expertise of the IDP with respect to controlling TB in occupational settings. Insight into the type of impact the TB program is having can been found in a response to an outbreak of 19 cases of active TB in a homeless shelter in St. Louis, Missouri (Martin and

Coffey 2005). After engineering controls (upper air and in-duct UVGI), increased fresh air, and RDRP-recommended improved maintenance practices were instituted, the outbreak ended.

Assessment of Relevance

The work of the TB program is highly relevant. RDRP work on respirators and environmental controls is at the forefront of such work, and the program clearly is a leader in this area. The numerous documents cited above and the RDRP's obvious role in that work supports this conclusion.

Assessment of Impact

The beneficial impact of the RDRP's TB work is very high. While it is not possible to quantify the extent to which the RDRP's work has reduced TB cases in occupational settings, on the basis of its work with engineering controls and respirators and the guidelines to which it has contributed, there can be little doubt that the program has substantially contributed to preventing TB transmission in a number of occupational settings (e.g., health care, correctional, and detention facilities, as noted previously). One threat to ongoing impact assessment is the lack of a formal occupational survey for TB and the RDRP's dependence on other CDC programs for these data (addressed in detail below).

Engineering Controls and Respirator Technology

In the evidence package materials, this area is presented as part of the IDP's TB control program. However, since environmental controls and respirator technology (particularly the latter) have implications for almost all occupational exposure sources, the area is given its own section here for evaluation.

Engineering control research focuses on ventilation and UVGI and ventilation systems. The personal respirator program focuses on bioaerosol filter efficiency, filter reuse, and fit testing. Some details that are relevant to this section were presented in the section on TB; others are included in the sections on anthrax and emerging infectious diseases. This section presents selected details that can be expected to have generic applications to many infectious diseases in occupational settings, particularly in health care facilities.

Planning and Production Inputs

Much of the work for engineering controls and personal respirator technology is driven by need, reflected by requests to control the spread of microbes in occupational settings. In this sense, the work is extremely responsive to the needs of stakeholders (health care workers, workers in prisons and detention facilities, health care workers and coworkers exposed to a large number of newly arrived immigrants from regions where TB is endemic). Integral to this work are risk assessments related to specific exposures. For example, NIOSH-funded, extramural investigators developed methods to document the concentration and size distribution of viable cough-generated *M. tuberculosis* from patients with TB (Fennely et al. 2004). A sampling and detection method for TB based on polymerase chain reaction was developed and published in the *NIOSH Manual of Analytical Methods* (NIOSH 1998b) and allows for determining the concentration and aerodynamic size of airborne mycobacteria.

Activities, Outputs, and Outcomes

Many of these elements are described in the section on TB. In addition, the RDRP has provided guidance on respirator use in relation to bioterrorism threats (NIOSH 2002b) and emerging infections (e.g., SARS and avian influenza) (CDC 2004a,b). Almost all guidance documents in the latter areas include recommendations for personal protection that directly reflect RDRP investigations into personal and building environmental controls to minimize transmission and infection. The RDRP contracted with the U.S Army's Dugway Proving Ground to evaluate the protective efficacy of surgical masks and dust/mist, dust/fume/mist, and high-efficiency particulate air (powered and nonpowered) respirators against *Bacillus subtilis* and NaCl aerosol. This work showed that these devices were equally protective against biological and nonbiological aerosols. Additional tests included filter efficiency, microorganism survival, mold and fungal contamination, and contamination due to handling. This work confirmed that Part 84 filters could be safely reused (Johnson et al. 1998). The RDRP also documented that passing of some fit tests listed in the OSHA respirator standard (29 CFR 1910.134) with N95 filtering face pieces does not guarantee an acceptable level of protection in the workplace (Coffey et al. 2004, 2006). This work was disseminated in comments to the docket on the proposed OSHA rule on occupational exposure to TB (see NIOSH 1998c).

Assessment of Relevance

The engineering control work is extremely relevant to protection of workers and minimization of the contamination of work facilities by highly infectious and lethal microbial agents. The RDRP is a leader in this area, as evidenced by the inclusion of the outputs from this work in virtually all CDC guidance on the protection from and spread of infectious agents in the workplace. This relevance is particularly true in health care settings and for first responders to real or potential bioterrorism or outbreak events.

Assessment of Impacts

As was the case for TB, there is no way to quantify specifically the impacts that work on personal respirators and environmental controls have had on reductions in the occurrence of respiratory infections. Clearly, in the case of the anthrax contamination of the U.S. Postal Service (USPS) facilities and the Senate Office Building, the use of the outputs of this work were integral to the protection of responders and decontamination teams. It seems fair to say that direct quantitative assessment of the impacts will never be possible, since the counterfactual experiment (exposure of the same workers without protection) would be unethical and immoral.

Other Components of the Infectious Diseases Program Goal

The evidence package presented several other elements of the RDRP under the headings of "anthrax," "emerging infectious diseases," and "understanding mechanisms of occupational exposures on pulmonary susceptibility." These do not represent a coherent related set of programs, but they are presented together here for the following reasons: (1) Anthrax presented a unique event that likely contributed to the RDRP's capacity to handle such events but does not represent an ongoing research program apart from its work on respirators and engineering controls. (2) A similar case can be made for SARS under the emerging infections. (3) Work on pandemic and avian influenza is largely preparatory and also depends heavily on work related to respirators and engineering controls. (4) The mechanisms component is not connected directly to other programs.

The anthrax program was developed in response to the anthrax attacks in 2001 that exposed postal workers to *Bacillus anthracis*. The IDP goal was to participate in the investigation of the exposures and to protect postal workers during the 2001 attack and in any future attacks. The closest that any statement comes to defining goals and objectives for any ongoing program is: "NIOSH and RDRP will maintain a state of readiness to respond within the context of CDC and PHS [Public Health Service] actions in the event of future attacks" (NIOSH 2006a, p. 196).

This rubric of emerging infectious diseases forms a core of current and future research activities and challenges for the IDP. The strategic goal is best expressed in a recent request for proposal (RFA-OH-06-002) "Prevention of Airborne Infections in Occupational Settings" (NIH 2006):

> Transmission of airborne infections to early responders, health care workers, and other occupational groups such as postal workers has become an important public health concern.... The objective of this RFA is to improve prevention of airborne infectious diseases in occupational settings.

Examples of the types of research of interest in the request for application (RFA) are as follows:

- Development of improved strategies for early identification and isolation of infectious cases.
- Development of improved approaches to detect and quantify airborne infectious agents and settled infectious agents with potential for re-aerosolization.
- Studies establishing exposure-response relationships for induction of disease by airborne infectious agents.
- Characterization of infectious aerosols generated by infected people, biological weapons systems, or other sources. Determination of infectious aerosol size distribution and impact of factors such as temperature, humidity, and ultraviolet irradiation on aerodynamic properties, viability, and infectivity of these aerosols. Elucidating factors that affect re-aerosolization of settled agents. Developing approaches to predicting the relative importance of airborne and contact disease transmission.
- Engineering controls such as optimization of ventilation and other building characteristics, and UVGI.
- Issues in the use of respirators to prevent transmission of airborne infectious diseases such as fit testing methods and interpretation, respirator selection, and innovative approaches to enable longer use or reuse of respirators when they are in short supply.

A section on histoplasmosis was provided in the evidence package (NIOSH 2006a). This program appears to be of interest historically since "no new RDRP research is planned or anticipated at this time...." Ample documentation was provided about the success of the RDRP efforts to protect workers in environments were the fungus is common and where outbreaks represent a threat. The evidence package states further that "we intend to continue dissemination of relevant information to prevent outbreaks of occupational histoplasmosis by maintaining awareness.... [I]nitiatives would only be undertaken should recurrence or worsening of the problem occur as documented by surveillance or disease outbreaks." However, in the absence of an operational definition of "awareness" and the documented lack of adequate resources for specific surveillance activities, the inevitable conclusion is that only the emergence of an outbreak or notification of suspicion of a problem by employees or management will trigger further activity in this area. This is far from satisfactory, but it is difficult to see how the RDRP could justify allocating scarce resources to a more proactive approach. No further comments are made about this program in the following sections.

Planning and Production Inputs

Anthrax

The RDRP played a major role in the national effort to protect workers from the threat of anthrax-contaminated mail by sending personnel to staff the CDC and Department of Health and Human Services Emergency Operations Centers and to the six locations where workers were at risk. The RDRP developed sampling procedures, recommended effective interim protective measures, safeguarded workers who decontaminated affected workplaces, assessed the effectiveness of decontamination, and disseminated information on protection. Sampling technologies were used in innovative ways, such as adapting a vacuuming technique using an "allergy sock," a method originally developed to measure allergens, to provide a more sensitive, comprehensive way to collect anthrax samples at large postal facilities.

The RDRP and partner agencies provided information to USPS managers, workers, and unions to help assess exposures at postal facilities and on Capitol Hill. The RDRP sampled numerous sites where contamination was known or suspected. The RDRP provided technical assistance on workplace sampling procedures and personal protective equipment for decontamination workers to aid in the EPA's clean-up of contaminated government buildings and worked with the EPA and others to determine when remediated congressional buildings were ready for reoccupancy by performing surface sampling after the clean-up and decontamination were completed.

Several ongoing projects are listed with respect to developing better understanding of how mail becomes cross-contaminated and of spore migration patterns in buildings; however, their relation to the "readiness" goal is not obvious. A more obvious realization of the readiness goal is a collaboration with Battelle Corporation to develop a database for anthrax sampling methods (collection and analytic methods) that is being made available to the public.

No specific NIOSH extramural funds have been set aside since 2001 for anthrax-specific research activities. However, as described above, a more general RFA was issued in 2006 that would permit such research.

Emerging Infectious Diseases (SARS, Pandemic, and Avian Influenza)

With respect to SARS, these activities are listed in Table 2-4. Of particular note is the RDRP's support role for the CDC's assistance (in the form of on-site investigations) to Canada and Taiwan during the SARS outbreaks in those countries. The RDRP also contributed staff to the CDC's airport operations related to the arrival of international flights in U.S. airports. The IDP contributed to two comprehensive CDC guidance publications that contained specific reference to transmission in occupational settings (CDC 2004a; NIOSH 2007b). An additional guidance document was prepared specifically for flight crews on commercial airlines (CDC 2004b).

With respect to influenza, the RDRP functions in the context of the CDC's overall program to deal with these issues. The IDP is taking the lead in workplace safety and industrial hygiene issues in several work groups: the nonpharmaceutical intervention work group, the working group for avian influenza (nonpandemic), respirator use for pandemic influenza, and technical assistance to OSHA in collaboration with the OSHA medical director. The IDP presents information on emerging issues to appropriate groups, such as the poultry industry (NIOSH 2006b). Therefore, in this context the CDC can be viewed as a major stakeholder to which the RDRP has been responsive.

Understanding the Effects and Mechanisms of Occupational Exposures and Pulmonary Susceptibility

Under the overall strategic goal of preventing and reducing occupational respiratory infectious diseases, the intermediate goal relevant to this work is to protect workers from occupational exposures making them susceptible to respiratory infections. NIOSH states that "Because this work is at the more basic end of the research to practice spectrum, its intermediate outcomes relate to influencing the thoughts and actions of others" (NIOSH 2006a). The committee queried NIOSH about whether this research is appropriate when other agencies (e.g., the National Institute of Allergy and Infectious Diseases [NIAID]) carry out similar research and what is unique to the occupational environment that requires their involvement in mechanistic research. NIOSH replied (NIOSH 2006b) that they are the only federal agency mandated to conduct occupational safety and health research, that their mechanistic research is consistent with that role, and that it complements the activities of other agencies. Further, the RDRP's mechanistic studies differ from those of NIAID, which focus more on elucidating mechanisms of host defense against microorganisms and developing clinical responses such as vaccines and treatments, and the research is within the NIOSH mission as it studies an important aspect of the toxicologic effects of occupational chemicals and exposures. The evaluation committee agreed that this is a reasonable and compelling rationale for the existence of this program.

Activities Outputs and Outcomes

Anthrax

Specific responses to stakeholders are listed in the preceding section and constitute important activities. A number of specific outputs and transfers have resulted from these activities (Table 2-4). A substantial number of technical papers, books and book chapters, numbered NIOSH publications, control technology, and HHE reports and presentations at conferences and meetings were produced. Worker fact sheets (paper and web-based) (NIOSH 2002c; CDC 2007b) were produced for use by first responders, workers, and investigators. Recommendations were published in conjunction within the larger CDC effort (CDC 2001), with specific IDP contributions related to engineering controls and protective equipment for personnel. The IDP contributed to the NIOSH publication related to procedures for collecting and analyzing samples to detect *B. anthracis* (NIOSH 2002c; Teshale et al. 2002; CDC 2007a). IDP staff also helped to train the Federal Bureau of Investigation, U.S. Coast Guard, independent contractors, and other personnel in appropriate procedures for anthrax decontamination and to determine when decontaminated buildings are safe for reoccupancy. Symptom surveys among exposed workers and air monitoring also were conducted followed by release of relevant reports (Hall and Bernard 2002; Hall and Hess 2002; Hall et al. 2002). One of the most important outputs to which the IDP contributed was the 2002 emergency preparedness plan developed in collaboration with the USPS (USPS 2002) and its associated biological detection system (USPS 2005) and the development and installation (four distribution centers) of new ventilation and filtration systems to capture releases of bioterrorism agents during mail processing. Guidance documents for protecting buildings from future bioterrorism attacks and filtration and air cleaning systems to protect buildings were also released (NIOSH 2002b).

Emerging Infectious Diseases

With respect to SARS, the main outputs have been guidance documents (CDC 2004a,b; NIOSH 2007b). In addition, the evidence package lists seven URLs on the CDC website to which it has made contributions with respect to workplace protection (e.g., respirators, schools, air transport, health care). RDRP staff published one peer-reviewed (Christian et al. 2004) paper and contributed to a CDC document that provides specific guidance for flight crews on commercial aircraft (CDC 2004b).

The RDRP has been highly productive with respect to planning for outbreaks of pandemic or avian influenza. As described by NIOSH (2006a), a number of specific guidance documents have been developed:

- Guidelines for Protecting Healthcare Workers Caring for Patients with Avian Influenza (CDC 2004c).
- Respiratory Hygiene/Cough Etiquette in Healthcare Settings (CDC 2003).
- Guidance on occupational health and safety aspects of avian influenza. (CDC 2007b).
- The NIOSH topic page covers occupational aspects of avian influenza (NIOSH 2007c).
- Pandemic Preparedness Checklist.
- Toolkit for Business Pandemic Flu Planning.
- CDCIDEOC Plans and Exercises: The CDC CONPLAN (internal operations plan for pandemic influenza) was released in February 2006.
- HHS Worker/Employee Protection Guidelines and Policy.

Extramural RDRP research on the efficacy of N95 respirators against virus-sized particles (University of Cincinnati) was completed (Lee et al. 2005). Various technical presentations, meeting disseminations, NIOSH interagency work group meetings, and meetings with the FDA have occurred. The Institute of Medicine was contracted to review the availability of surgical masks in the event of a pandemic and various use characteristics of personal respirators. The report was completed in 2006 (IOM 2006). Research efforts are also under way to assess the reusability of disposable filtering face piece respirators to protect against infectious aerosols (71 Fed. Reg. 56151[2006]).

The IDP conducted a review of the Department of Homeland Security (Coast Guard) internal pandemic flu plan with regard to protecting department personnel. The IDP is working with the Department of Homeland Security on this issue.

The most notable intermediate outcome relates to meetings with respirator manufacturers to address concern about the demand for NIOSH-certified N95 disposable respirators. RDRP scientists played a role in the effort to convince WHO to revise the posted recommendations for respiratory protection by noting which other disposable respirators are comparable to the N95.

Understanding the Effects and Mechanisms of Occupational Exposures and Pulmonary Susceptibility

Activities, outputs, and outcomes are summarized in Table 2-4. Several findings from RDRP studies can be highlighted. Studies have found that exposing rats to diesel exhaust particulate matter but not carbon black (the carbonaceous core of diesel exhaust particulate matter without the organic compounds adsorbed onto it) significantly increased the susceptibility of their lungs to infection with *Listeria monocytogenes* (Antonini et al. 2001a). Other studies have addressed lung clearance function and the pulmonary clearance of *L. monocytogenes* after exposure to diesel exhaust particles, different welding fumes, and silica (Antonini et al. 2001b, 2004; Yang et al. 2001). Although *L. monocytogenes* is not an important pathogen in the occupational setting, the choice of it as a test organism is justified because host defense responses after intrapulmonary deposition of the organism have been well characterized and involve an initial antigen-nonspecific host defense mediated by alveolar macrophages, followed by development of adaptive cell-mediated immune responses in which T cells network with alveolar macrophages and other cells for host defense. RDRP scientists have developed an innovative, automated, robotic welding fume generator and inhalation exposure system for use with laboratory animals to simulate real workplace exposures (Antonini et al. 2006). RDRP scientists have determined the mechanisms of how workers such as welders in construction and boilermakers who have inhaled metal-containing particles of mixed composition become more susceptible to infection than the general population. The RDRP research on occupational respiratory diseases in welders won the 2006 American Welding Society Safety and Health Award.

The major intermediate outcome related to partnerships with other federal agencies involves an interagency agreement with the Department of Defense (DOD) entitled, "Effects of Mixed Dusts on Asthma and Pulmonary Infectivity." Results from this work may contribute to DOD's development of an exposure standard for diesel exhaust particulate matter. A second project with DOD involves pulmonary and central nervous system toxicity associated with welding fumes. The relevance of this project to the IDP cannot be ascertained from the documentation materials supplied to the committee. The NTP has funded the RDRP to study immune response in the lungs associated with exposure to welding fumes. The NTP plans to use the RDRP model for welding fume generation and inhalation for its chronic neurotoxicity and carcinogenicity studies in addition to those related to immune function in the lung.

Assessment of Relevance

Anthrax

Unquestionably, the work carried out by the RDRP was very relevant to controlling the spread of anthrax within USPS facilities and the Senate Office Building and the protection of workers during decontamination work. The activities of research and testing of respirators and engineering controls were an important factor in the effectiveness of the RDRP's efforts. The CDC advisories to which the

RDRP contributed are highly relevant to worker protection should any such contamination occur in the future.

Emerging Infectious Diseases

The relevance of the work related to SARS and influenza can be considered together. Again, this work is highly relevant and rests on the RDRP's capacity to develop rapid methods of detection and worker protection through respirators and engineering controls for workplace facilities.

Understanding the Effects and Mechanisms of Occupational Exposures and Pulmonary Susceptibility

As described above, the committee determined that the RDRP's research is distinct from that of other federal agencies. This focus area is relevant and will support the missions of other federal agencies as evidenced by the previously described interagency agreements with DOD. What appears to be lacking is a cogent strategy for determining which specific elements of host defense are going to be addressed, in what priority, and in relation to which exposures. Although some future projects are identified, the RDRP does not state how the program has and will complement, rather than duplicate, research being done by other federal agencies on similar exposures (e.g., the EPA-funded research on the toxicity and health effects of diesel particulate matter).[3] Thus, to ensure the relevance of this component, there is a need to develop a more explicit overall program of research and to clarify its relation to studies being carried out by other federal agencies.

Assessment of Impacts

Anthrax

It is nearly impossible to disentangle the specific role of the IDP or, for that matter, any other agency involved in limiting the spread of *B. anthracis* and, thereby, decreasing the risk of infection and death in exposed workers and investigators. However, insofar as IDP plays a major role in the development and dissemination of personal and building respiratory protection systems, it is fair to credit the IDP with helping to contain what could have been a larger disaster with greater cost of human lives.

[3]One example of a more specific line of inquiry might be to investigate the potential mechanisms underlying the increased risk of bacterial pneumonia among welders and how it might relate to iron particulates and facilitating infectivity or virulence. Similarly, mechanistic questions about silica's promotion of atypical mycobacterial disease remain to be fully elucidated.

Emerging Infectious Diseases

Similarly, with regard to SARS, it is nearly impossible to disentangle the specific role of the IDP or, for that matter, any other agency involved in limiting the spread of SARS and, thereby, decreasing the risk of infection and death among the public and in exposed health care workers and workers in the airline industry. However, insofar as the IDP plays a major role in the development and dissemination of important guidance documents for preventing the transmission of the SARS coronavirus, it is fair to credit the IDP with helping to contain what could have been a larger disaster with greater cost of human lives.

Since there has not been an influenza pandemic or a large outbreak of avian influenza in the United States or other parts of the world, there are no outcomes by which this program can be evaluated. There are, however, some important challenges related to likely future impacts noted by the committee. Successful completion of these challenges could be used as appropriate impacts in advance of any outbreaks. RFA-OH-06-002 (NIH 2006) has already been mentioned to solicit extramural research in a host of areas related to emerging infections, including the identification, characterization, and control of infectious agents.

This RFA represents an important step to engage extramural researchers in the IDP emerging infections program and appears to be an appropriate prioritization of extramural funding in the face of limited budgetary resources. However, the evidence package does not explicitly indicate how the RDRP intends to integrate the research generated.

The IDP is to be commended for its work on personal respirators and engineering controls relevant to previous transmission of infectious agents to workers. The RDRP through the IDP has played a leading role in the CDC's efforts to evaluate and disseminate these technologies.

A major challenge to the IDP is the limited availability of BSL-3 laboratory space in which to carry out studies of aerobiology, transmission potential, exposure assessment methods, engineering controls, and personal protective equipment. Efforts are under way through the CDC to establish an appropriate facility to be located on the National Biodefense Research Campus in Fort Detrick, Maryland.

Understanding the Effects and Mechanisms of Occupational Exposures and Pulmonary Susceptibility

No overall statement of possible end outcomes was provided in the documentation materials in response to the committee's queries about this program. Outcomes for specific research projects are noted, but they are not integrated into a coherent set of program outcome criteria that could be used to evaluate impact. However, insofar as the studies have improved our understanding of susceptibility and immune mechanisms, it is fair to credit the RDRP with contributing to our understanding of lung defenses related to a variety of respiratory infections.

STRATEGIC GOAL 4: PREVENT AND REDUCE WORK-RELATED RESPIRATORY MALIGNANCIES

Introduction

Within the overall RDRP mission the goal of this component is to "prevent and reduce work-related respiratory malignancies."

Before NORA, the strategic plan for preventing occupational lung diseases did not include lung cancer as a specific disease target. While occupational cancer was a general NIOSH priority, this work was performed by investigators in the Division of Surveillance, Field Studies and Hazard Evaluation. When the NORA priorities were established, attention to respiratory malignancies was located in the priority Cancer Research Methods. Assessments of the overall cancer burden, at that time, drew attention to the estimated annual occupational lung cancer burden of 9,000-10,000 men and 900-1,900 women. Half of this burden was attributed to asbestos-related bronchogenic carcinoma. Furthermore, a much larger number of workers were noted to be or to have been exposed to known human respiratory carcinogens, not to mention the millions exposed to substances that are probable or possible human carcinogens, some of which carry risk specific to the respiratory tract (Toraason et al. 2004).

To address the RDRP subgoal of preventing and reducing work-related respiratory malignancies, the RDRP selected the following targets as most relevant.

- Determine occupation etiologies of lung cancer.
- Reduce metal-induced lung cancer (hexavalent chromium).
- Prevent and reduce silica-induced lung cancer.
- Prevent and reduce lung cancer induced by diesel engine exhaust.
- Produce lung cancer diagnostic tools.

Our comments in this chapter address the planning and production inputs for all five of these goals, followed by a review of the activities and outputs for the same issues. We follow these reviews by assessing the relevance of these materials and the impacts of the RDRP's lung cancer research effort.

Planning and Production Inputs

Determination of Occupation Etiologies of Lung Cancer

The relevance of the overall goal of prevention and reduction of work-related respiratory malignancies is well established and appropriately documented in the RDRP evidence package. In particular, NIOSH references the CDC's Healthy People 2010 goal to reduce further the nation's cancer burden (CDC 2000), the primary importance of lung cancer due to its rank as the leading cause of deaths from cancer in the United States, the substantive attributable fraction for lung cancer due to occupational exposures (Steenland et al. 1996), and the large number of workers in the United States thought to be exposed to known or suspected carcinogens at work. RDRP scientists have been key authors or contributors of many publications that have established these estimates, which in turn provided the rationale to make preventing occupationally related lung cancer a priority for the program.

The RDRP has participated in several planning processes in reference to occupational cancers including lung cancer. A comprehensive occupational cancer-planning process was conducted by the Industry Wide Studies Branch (Ward et al. 1993) and later through the NORA committee process (Ward et al. 2003). The latter report is very comprehensive, although it focuses on priorities for research methods rather than specific etiologic research needs. An earlier planning document on the prevention of occupational lung diseases in general limited its focus on cancer to those associated with asbestos (NIOSH 1986a).

These planning processes provide a basis for ensuring that the research activities undertaken by the RDRP are relevant to the overall NIOSH research agenda; however, the linkage between the planning documents and the five subgoals and the myriad specific activities is not clear. In contrast, the project estimating the attributable fraction and mortality rate in the United States (Steenland et al. 1996) and worldwide (Driscoll et al. 2005b) is highly relevant to the mission.

The RDRP mentions significant focused efforts on lung carcinogenic effects of cadmium, ethylene oxide, chrysotile asbestos, beryllium, uranium, and radiation among exposed workers as well as an evaluation of cancer risk among construction painters. The specific program inputs and activities for these studies, however, are not addressed in detail.

Including specific respiratory cancers as a part of the RDRP portfolio rather than in a program addressing occupational cancers in general appears somewhat arbitrary and awkward. Presumably, their inclusion in the RDRP stems from the major involvement of Division of Respiratory Disease Studies personnel in these programs (e.g., silica and diesel exhaust) rather than from an explicit strategic decision. In fact, with the exception of these two specific topics, the respiratory cancers are included under the general occupational carcinogens planning documents and NORA agenda. Because of the commonality of methods and expertise used for studying occupational cancers, assigning respiratory cancer into a separate program area may create counterproductive divisions among researchers. On the other hand, the management of respiratory malignancies in the new matrix structure of the RDRP may ensure better communication among units in different cities within the reorganized RDRP programs.

Although the choice of the three exposure-specific subgoals—chromium, silica, and diesel exhaust—appears to be highly relevant, the rationale for choosing these particular risks over other ongoing lung carcinogen exposures, or other non-respiratory occupational carcinogens, is not addressed.

Reduce Metal-Induced Lung Cancer (Hexavalent Chromium)

Hexavalent chromium (chromium VI) has been long recognized as a lung carcinogen and NIOSH addressed the problem in an early criteria document in 1975 (NIOSH 1975). Despite this recognition, OSHA did not begin to regulate chromium VI exposure as a carcinogen until 2001. As NIOSH is expected to respond to an OSHA need for the best evidence base to guide its regulatory actions, the RDRP placed a priority on addressing OSHA's need for additional information and technology in support of a possible revision of its existing standard for exposure to chromium VI. The RDRP developed projects on quantitative epidemiology and risk assessment and the development of sensitive and rapid and field-portable detection methodology for chromium VI. In both areas, the RDRP's contributions demonstrate the special expertise available within the program. RDRP epidemiologists and biostatisticians were able to exploit established cohorts with well-characterized chromium exposures to explore quantitative aspects of the exposure-response relationships and to use them to model risk in exposed populations. RDRP scientists also overcame a significant hurdle in relation to field monitoring for exposures to chromium VI. The problem existed for settings where exposure to chromium VI is possible but levels are low—for example, in the construction industry. The RDRP succeeded in developing environmental monitoring technologies to determine trace levels of chromium VI with sufficient sensitivity for detection where low-level exposures occur.

Prevent and Reduce Silica-Induced Lung Cancer

In the case of silica-induced lung cancer, OSHA has sought guidance from NIOSH for regulation of exposures to silica but has been inconsistent in developing its regulatory approach to silica. However, the RDRP has recognized the potential importance of the issue because of the ubiquity of the exposure and the mounting evidence for carcinogenicity. In addition, the RDRP's long-standing interest in silica exposures in a variety of relatively high exposure settings provided the basis for addressing the emerging recognition of the association between silica and lung cancer. NIOSH responded with two major epidemiologic studies that helped provide evidence on this issue. The first, a study of gold miners, indicated excess risk for silica-exposed workers but exposure information was not sufficiently detailed to document an exposure-response relationship. The second was a study of Vermont granite workers where, with higher-quality exposure information, the excess risk was shown to increase with exposure level (Steenland and Brown 1995; Attfield and Costello 2004). NIOSH's ability to respond was clearly related to long-standing familiarity with both silica exposures and cancer epidemiology expertise.

Prevent and Reduce Lung Cancer Induced by Diesel Engine Exhaust

The issue of the association of diesel exposure with lung cancer risk provides an especially revealing picture of the RDRP's relevance and impact. NIOSH first published a document assessing diesel exhaust as a probable human carcinogen in 1988 (CDC 1988). In a sense, this current intelligence bulletin served as a planning document and provided evidence for the importance of the issue. Similar conclusions were published by IARC (1989), WHO (1996), NTP (1998), California Environmental Protection Agency (CalEPA 1998), and EPA (2002). In 1997, the National Cancer Institute and NIOSH agreed on an interagency project protocol for two studies—a cohort mortality study with a nested case-control study of lung cancer and diesel exhaust among nonmetal miners. These studies are close to completion (see below).

MSHA promulgated comprehensive standards to control diesel exhaust in underground mines in 2001. In addition to contributing to the development of these standards, the RDRP has also responded with a series of projects and studies to comprehensively respond to several challenges. The studies include detailed quantitative epidemiologic and risk assessment studies; development of air-sampling technologies suitable for monitoring diesel exhaust in the presence of other carbon-containing particulate matter, such as that found in underground coal mines; and studies of alternative approaches to control technologies for diesel exhaust. There are significant nondiesel sources of carbon particles in mines that interfere with attempts to measure the concentrations of diesel particulate matter. As a result, NIOSH has worked with the mining community to develop methods of measuring tail-pipe emissions of diesel engines in coal mines and of measuring environmental concentrations of diesel particulate matter in other (e.g., metal and nonmetal) mines that have led to exposure regulations (see http://www.cdc.gov/niosh/nas/mining/potentialintermediateoutcome5.htm).

Produce Lung Cancer Diagnostic Tools

The fifth subgoal addresses lung cancer diagnostic tools. The rationale for these activities is based on the significance of lung cancer in the general population and the importance of early detection and staging in secondary prevention. The linkage of these activities to the RDRP planning processes and stakeholder input is unclear. In fact, the current planning document on priorities for developing research methods in occupational cancer (Ward et al. 2003) places an appropriate focus on workplace exposure and workforce approaches but makes no mention of the priority for developing lung cancer diagnostic markers. For example, the document addresses the importance of intervention studies among high-risk cohorts but does not mention early detection methods as an important area for development. Although RDRP scientists may make significant contributions to this area of lung cancer research, NIOSH's mission to address work-related injury and illness does not provide a clear rationale for this activity. Other research agencies may be better suited to these issues, permitting NIOSH to allocate these resources to laboratory explorations directly related to cancer detection even if the tool could be applied in cohorts with elevated risk of occupationally related cancer.

Activities and Outputs

The impact of the RDRP work under the stated subgoals has been evaluated in terms of the transfer of the outputs to the broad audience of potential users or partners as well as in terms of evidence that intermediate or end outcomes have been accomplished. For each of the five goals associated with the prevention of occupational respiratory malignancy (determination of occupational etiologies; reduction of metal-induced lung cancer, silica-induced lung cancer, and diesel exhaust-induced lung cancer; and development of diagnostic tools) consideration was given to evidence of impact in terms of intermediate end points and current program activities. Not considered here are the major contributions by NIOSH to the study of lung cancer related to asbestos exposures and to radon daughters among uranium miners. This work is no longer considered high priority and, for the most part, studies of these risks were completed before the original NORA.

While the end impact of each of these goals is a reduction in the incidence of or mortality from respiratory cancer, the long latency between exposure and cancer onset makes it unrealistic to gather adequate direct evidence of impact. Therefore, the committee considers evidence of impact on intermediate outcomes as providing adequate support for an impact on morbidity and mortality.

Determine Occupational Etiologies of Lung Cancer

During the past 10 years, RDRP scientists have published at least 200 peer-reviewed articles and book chapters (127 since 2001) on occupationally related lung cancer. More than 90% of those published after 2000 were peer-reviewed journal articles. They have contributed valuable scientific research findings to the scientific literature. The RDRP's engagement in the transfer of these published research findings has been primarily through the standard routes of public presentations at scientific and professional meetings. There are some notable specific efforts, including testimony at an executive branch effort in 1994 to assess the role of government agencies in the research mission of the National Cancer Program (L. L. Fine, NIOSH, testimony before the President's Cancer Panel, January 31, 1994). At that time, the director of the Division of Surveillance Hazard Evaluation and Field Studies was primarily responsible for the strategic planning and priority setting concerning occupational cancer.

Although not a stated priority effort, RDRP scientists contributed to the efforts to control secondhand cigarette smoke both through solicited commentary in the June 2006 Surgeon General's Report on the Health Consequences of Second-hand Smoke" (DHHS 2006) and through the development of a number of fact sheets based on the content of the report for dissemination to a wide variety of stakeholders and customers: legislators, employers, unions, physicians, parents, and others. Although this risk is not only an occupational risk, it certainly is an important risk in work environments.

A long-standing and remarkable program of the RDRP is its effort to directly notify individual workers who were part of cancer epidemiology studies of the study findings; this is in accordance with the NIOSH Worker Notification Program (NIOSH 2007d). The RDRP estimates that it has successfully notified more than 48,900 workers about lung cancer hazards (NIOSH 2006a).

In recent years, regulatory agencies have taken much less advantage of NIOSH research findings than was the case in the early years of NIOSH's existence. Two important exceptions are the OSHA cadmium standard (29 CFR 1910.1027) issued in 1992, which relied on RDRP findings especially with respect to risk assessment (58 Fed Reg. 21778 [1993]), and the final OSHA rule on hexavalent chromium, which relied on both the epidemiologic research and the risk assessment findings of NIOSH. Other agencies, however, have made good use of the NIOSH research to affect public policy, as evidenced by the California Air Resources Board action on diesel particulate matter (see below).

RDRP studies and risk assessments have contributed to evidence used by IARC in classifying agents as carcinogenic, including beryllium, cadmium, silica, and chromium (IARC 1993). IARC has included evidence on excess risk for lung cancer from the RDRP when categorizing 13 agents as Group 1 (carcinogenic to humans), 10 as Group 2A or 2B (probably carcinogenic [2A] or possibly carcinogenic [2B]), and 1 as Group 3 (not classifiable with respect to carcinogenicity). Since IARC does not identify the relative importance of individual studies, it is not possible to measure how large an impact specific NIOSH studies had in affecting the IARC determination. Nonetheless, several of the IARC determinations include a substantial number of NIOSH references, indicating they were a major factor. In the case of beryllium, however, it is clear that NIOSH's research was seminal in the determination that this metal is a human carcinogen. For the most part, the major occupational carcinogen classifications by IARC classification occurred before this century; however, NIOSH continues to study the agents in an effort to improve understanding of detection and control.

Reduce Metal-Induced Lung Cancer (Hexavalent Chromium)

The Occupational Safety and Health Act of 1970 (29 USC 671) authorized NIOSH to "develop and establish recommended occupational safety and health standards" [Section 22(c)(1)] as well as "to conduct such research and experimental programs as he determines are necessary for the development of criteria for new and improved occupational safety and health standards, and . . . after consideration of the results of such research and experimental programs make recommendations concerning new or improved occupational safety and health standards" [Section 22(d)].

The RDRP risk assessment contributed significantly to OSHA rulemaking that resulted in a revised chromium VI (hexavalent chromium) standard promulgated in early 2006 (71 Fed. Reg. 10099[2006]). The RDRP risk assessment estimated that exposure corresponding to a lifetime risk of 1/1,000 was 0.2 $\mu g/m^3$ (Park et al. 2004). Furthermore, the assessment found no evidence to support a threshold in the intensity effects of chromium VI and no evidence to suggest saturation of defense mechanisms (such as extracellular reduction of chromium VI deposited on the lung epithelium) (Park and Stayner 2006). While OSHA found the NIOSH risk assessment compelling, the agency determined that considerations of technological and economic feasibility led them to set the PEL at 5 $\mu g/m^3$, which is associated with an excess lifetime risk per 1,000 exposed persons of 10-45 lung cancer deaths compared with 2-9 cancers per 1,000 per lifetime at the NIOSH recommended PEL of 1 $\mu g/m^3$. The seven RDRP publications relating to chromium VI are listed elsewhere (Wang et al. 1997; Boiano et al. 2000; Ding et al. 2000; Ashley et al. 2003; Park et al. 2004, 2006; Park and Stayner 2006).

The work of RDRP scientists is cited in OSHA's "Final Rule on Occupational Exposure to Hexavalent Chromium" along with the work of others (71 Fed. Reg. 10099 [2006]). OSHA had originally argued against including the construction industry within the scope of the proposed rule, citing the lack of field-portable analytic capability. As part of the effort to assist OSHA in improving the feasibility of controlling exposures to hexavalent chromium, NIOSH developed and patented (U.S. Patent 6,808,931) a field-portable, rapid chromium VI detection method [NMAM (Method 7703)]. The field technique allows for the measurement of chromium VI at levels well below the new OSHA PEL and thus permits assessment in occupations that are not normally monitored, such as construction (Boiano et al. 2000; Wang and Ashley 2004). Largely as a result of this new method, OSHA ultimately included construction within the scope of the final rule that was promulgated. The U.S. Air Force is already using the RDRP-developed field-portable rapid chromium VI detection method in aircraft painting/maintenance operations (Aizenberg et al. 2000). NIOSH anticipates that applications of the field chromium VI method will become more widespread now that the new OSHA rule has been promulgated.

RDRP scientists continue to be funded by OSHA to investigate some of the mechanistic issues that might prove relevant to chromium VI risk management and permit a more nuanced regulation of chromium VI exposures. These efforts are now laboratory based; the study of mechanisms of carcinogenesis and the role of genetic polymorphism are also areas of considerable interest.

Silica-Induced Lung Cancer

Publications from the RDRP on lung cancer among workers exposed to silica (Costello and Graham 1988; Steenland and Brown 1995; Attfield and Costello 2004) have been influential in the ongoing debate on the carcinogenicity of silica dust. They were among the 10 most-influential studies for its evaluation of silica dust in 1997 when IARC designated crystalline silica as a Group I carcinogen (IARC 1997). These publications continue to inform discussion about silica-related cancer risk and were featured and discussed in an international workshop organized by a European industry group in 2005.

Studies of silica-associated oxidative DNA damage, gene activation, and carcinogenesis in a p53-deficient mouse model are ongoing (see Chapter 3). Given the extensive work NIOSH has conducted on silica carcinogenesis, it may be reasonable to ask if the work could be extended to address the potential carcinogenicity of other fibrogenic dusts.

Lung Cancer Induced by Diesel Engine Exhaust

NIOSH's responsibility to MSHA is similar to its responsibility to OSHA—undertaking research and developing proposals for health standards for submission to the assistant secretary. Several developments regarding a MSHA rulemaking on diesel exhaust exposures in underground mines have relied on products of the work of the RDRP. MSHA does not have sufficient in-house scientific personnel to undertake such studies on its own.

RDRP scientists published two key studies that were important to MSHA in their risk assessment for their final rulemaking for diesel exhaust. NIOSH contributed one major epidemiologic study to the evidence base for associating diesel particulate matter with lung cancer (Steenland et al. 1998). In addition, NIOSH's risk assessment paper (Stayner et al. 1998) provided essential assistance to MSHA in arriving at their conclusion about preventable lung cancer risk.

As a result of research by the RDRP, MSHA's final rule changed the interim standard's surrogate for measuring diesel exhaust particulate matter in metal/nonmetal mines from total carbon to elemental carbon (Birch and Noll 2004; 71 Fed. Reg. 33387 [2006]).

Similarly, RDRP research documenting the likely success when available control technologies for diesel particulate matter were applied in metal/nonmetal mines was important in gaining acceptance of the new MSHA rule. The research results, in addition to being published, have been shared at two workshops and partnership meetings attended by representatives of industry, labor, and government (Bugarski and Schnakenberg 2001, 2003; Schnakenberg and Bugarski 2002, 2004; Bugarski 2004; Bugarski et al. 2004a, 2005, 2006; Schnakenberg et al. 2005).

On January 19, 2001, MSHA promulgated final rules to control exposures to diesel exhaust in underground coal mines (66 Fed. Reg. 5526 [2001]; 30 CFR 72) and in underground metal and nonmetal mines (66 Fed. Reg. 5706 [2001]; 30 CFR 57).[4]

The RDRP, MSHA, and a private company (SKC Inc.) engaged in research on a new commercially available sampler. Keeping industry and labor groups informed throughout development of the sampler facilitated both the transfer and acceptance of the new technology, which is now the MSHA-required sampler for

[4]In coal mines and metal and nonmetal mines, there are significant nondiesel sources of carbon particulate matter that interferes with efforts to measure the concentration of diesel particulate matter, most of which is elemental carbon. This problem is solved in coal mines by measuring tailpipe emissions of diesel engines and, in metal and nonmetal mines, by measuring environmental concentrations. NIOSH has developed methods for making both types of measurements and for developing technology for achieving these limits. A lingering question is whether these two very different approaches provide equally safe environments.

measuring diesel particulate matter in metal/nonmetal mines (Noll et al. 2005). MSHA relied on RDRP research (Bugarski et al. 2004b) as part of the evidence that available technologies are capable of reducing the amount of particulate matter in diesel exhaust.

In a separate arena, NIOSH contributed important evidence to the California Air Resources Board proceedings that led to the decision to list diesel particulate matter as a "toxic air contaminant." The same two papers that MSHA relied on for its action were considered relevant in California as well.

NIOSH and the National Cancer Institute are now in the process of completing a major epidemiologic study of diesel exhaust and mortality (particularly from lung cancer). The study includes a large cohort mortality study, a nested case-control study, and an extensive effort to retrospectively estimate job-specific exposure over a wide range using historical records and new industrial hygiene surveys. The study seeks to characterize the quantitative relationships between diesel exhaust and cancer and, therefore, is focused on nonmetal miners, where exposure levels are probably the highest and there is little exposure to other lung carcinogens. It is reasonable to expect that this study will provide the most detailed and comprehensive assessment to date of lung cancer risk from exposures to diesel particulate matter.

The study's progress has been affected by intensive scrutiny and legal action by industry. Legal challenges to the process that initiated the study required NIOSH and the National Cancer Institute to repeat the protocol peer-review process from the beginning (Monforten 2006). This, too, was challenged. The consequence was substantial added delay in progress as well as a court order to require the government to refrain from publicly releasing all information submitted to the House Committee for a 90-day period. Work is proceeding and will be completed under the terms of this court order.

The MSHA final rule still presents assessment problems at the lower range of exposures due to technical limitations of the exposure assessment methodology. A key to addressing this is the RDRP's ongoing work on finding a reliable ratio between elemental carbon and total carbon at these measurement levels to allow MSHA to establish an elemental carbon surrogate for total carbon at these lower limits and thus enable MSHA to enforce the new exposure limit.

Research, expected to begin soon, is being planned by RDRP scientists to address long-term implementation issues relating to the MSHA diesel exhaust rule. NIOSH estimates that approximately 1.35 million workers are occupationally exposed to diesel exhaust particulate matter in about 80,000 workplaces in the United States (MMWR 1989b). Exposed workers outside of mining include railroad workers, bus garage workers, trucking company workers, forklift truck operators, firefighters, lumberjacks, tollbooth and parking garage attendants, and many professions that service or handle automobiles (car mechanics, professional drivers) (NTP 2005).

Lung Cancer Diagnostic Tools

The RDRP research on spectral karyotyping and comparative genomic hybridization has been applied to identifying genetic changes that occur during the general progression of lung cancer. In addition to published reports (Sargent et al. 2002; Yuan et al. 2004) a cooperative research and development agreement with Spectral Genomics, Inc., has resulted in development of a syntenic genomic chip and a patent is pending (2004-025124). This mouse-human chip will be used to allow direct comparisons between mouse and human lung adenocarcinomas.

This RDRP research has led to generalizable knowledge about the mechanisms of lung cancer that is valuable to other lung cancer researchers. The syntenic chip is the forerunner of a human chip that will be used to diagnose and stage lung adenocarcinomas in humans.

The RDRP believes that these findings can be applied to mouse and human studies to establish biomarkers for occupational exposure to lung carcinogens and the early detection and staging of lung cancer. The cooperative research and development agreement established with Spectral Genomics will assess the applicability of the observed gene expression and gene copy number changes detected in RDRP studies to aid in the early detection, diagnosis, and staging of lung cancer.

While the committee found evidence of the impact of the RDRP research effort in addressing the development of early markers for lung cancer starting with a mouse model, the potential for this work to lead to biomarkers of occupational exposures was not documented or explained. The research appears to the committee to be distant from addressing occupationally related respiratory malignancies and might equally well be carried out by the National Cancer Institute or the National Institute of Environmental Health Sciences, each of which has significantly more research resources and a mandate closer to this particular goal.

Assessment of Relevance

The RDRP has addressed several of the key occupational lung cancer risks in substantial and relatively comprehensive ways. The relevance of the work addressing the three carcinogen-specific exposure goals—silica, hexavalent chromium, and diesel exhaust—is unquestioned. These three carcinogens carry a significant risk, expose large occupational populations, and are highly amenable to control. In fact, the programs addressing these goals may have been motivated both by internal scientists' recognition of the problem and by external opportunities for having impact. Although the committee has no specific concerns related to the choice of these three exposures' programmatic priorities, the rationale for their prioritization over other possible important lung carcinogens is not evident.

NIOSH has also conducted very significant work on the contribution of occupational exposures to lung cancer risk in working populations. To some degree, the relevance of these efforts has been aided by substantial planning documents produced in 1993 (occupational cancer in general) and in 2003 as part of the NORA process addressing occupational cancer research methods. However, as stated previously, although there are some advantages for RDRP involvement in respiratory cancer, respiratory concerns should be included in a broader program addressing occupational cancers in general. The NORA team and the planning documents cited earlier addressed occupational cancers but did not separate lung cancer for attention by a different program. Furthermore, the degree to which information from surveillance activities, HHE/TA projects, or reports from the field contribute to the planning of priority areas for research on lung cancer is not clear.

While the relevance of early detection methods for lung cancer is not disputed, the relevance of this area of investigation to NIOSH's unique mission within the federal health research apparatus is questionable. A closer link between the biomedical research activities of the RDRP and the unique problems associated with occupational exposures, populations, or research methods is needed.

It is important to reiterate a key characteristic of the RDRP's three areas of exposure-specific subgoals. In each case, the RDRP has responded to a need recognized by either the scientific community or specific regulatory agencies, responding with multiple approaches to address the complexity of exposure and risk in the workplace. The ability to conduct quantitative epidemiology and risk assessment, while also addressing chemical and technological challenges of exposure assessment and exposure controls is an unusually effective approach for a scientific or governmental agency. The multidisciplinary and comprehensive approach to addressing risks in the workplace is a hallmark of the RDRP and NIOSH's contribution to public health.

Assessment of Impact

After review of the contributions presented by the RDRP and the input of stakeholders, there is evidence of solid and effective research that has had a direct impact on the control of occupationally related cancer risk in today's workplaces. There is current evidence that the research products have been of direct use both to OSHA and to MSHA in their standard-setting process as seen in the value of the RDRP risk assessments for both hexavalent chromium and for diesel exhaust particulate matter. The positive impact of the RDRP transfer efforts is also evident in the importance the agencies place on RDRP research that has been called for and effective in addressing complex issues related to methods to permit field assessment of low-level exposure to hexavalent chromium and relatively low-level exposure to diesel exhaust particulate matter. While the RDRP work on silica and respiratory malignancies has not penetrated as well into affecting agency rules and stakeholder practices, the reasons for this are beyond NIOSH's control.

Ongoing research continues to address some of the challenging problems related to occupational lung cancer risk. The RDRP has effectively engaged with stakeholders from both the workforce and the industry. This appears to be important in gaining acceptance of the research findings as well as adopting the methods and technology that were the subject of the research.

It is important to note the unusual and successful efforts the RDRP has directed at informing, individually, all subjects of cohort studies about the findings from these studies and what meaning it could have to each worker. This attention to communicating to occupationally exposed populations is a model for investigators both within and outside government.

The impact of the respiratory malignancies program is particularly strong in those specific areas where NIOSH applies its special expertise and mandate, comprehensively addressing specific carcinogenic exposures in the workplace. This strength is well demonstrated in the case of silica and chromium VI exposures. The work in diesel exposures has been completed, and it is expected to be integrated into this highly relevant exposure and risk assessment. The program components that address the contribution of occupational exposures to the burden of cancer in general are very significant. However, this contribution is not particular to respiratory cancers and it is unclear that the program is well served by separating these particular outcomes from other important occupational cancer risks. Finally, the committee questions the specific relevance to the NIOSH mission of the research directed at diagnostic tools for early detection of lung cancer. While this area of investigation is relevant to lung cancer in general, there is little advantage of having this work located at NIOSH. Focusing the early detection biomarkers on workplace-specific prevention efforts would make these efforts more relevant to NIOSH's mission and would increase the likelihood of the program's having a significant impact.

STRATEGIC GOAL 5: PREVENT RESPIRATORY AND OTHER DISEASES POTENTIALLY RESULTING FROM OCCUPATIONAL EXPOSURES TO NANOMATERIALS

Introduction

The overarching theme for the NIOSH Nanotechnology Research Program is to understand and prevent injuries and illnesses due to occupational exposure to nanomaterials in the workplace environment. This is a relatively new program that was implemented before any relationship was observed between nanotechnology and occupational lung disease. Thus, it is a proactive program rather than one that attempts to resolve problems after they have occurred. It includes specific concern with potential risks to the respiratory system as a natural extension of the previous research by NIOSH on the occupational hazards associated with inhaled dusts.

NIOSH has indicated the following specific goals for the program:

- To conduct research aimed at understanding and preventing work-related injuries and illnesses due to the use of nanotechnology products. This involves determining the toxicity of nanomaterials, identifying potential health outcomes from the use of these materials, and monitoring the health of individuals who work with these materials. NIOSH will also serve a major role in disseminating guidance information related to engineered nanomaterials.
- To conduct research to prevent work-related injuries and illnesses due to the application of nanotechnology products. Nanotechnology-based sensors and communication devices may help in handling emergencies and in empowering workers to take preventive steps for reducing the risk of injury. Nanotechnology-based fuel cells, lab-on-chip analyzers, and optoelectronic devices have the potential to be useful in making the workplace safe and healthful.
- To promote healthful workplaces through intervention, recommendation, and capacity building. This involves developing and evaluating engineering controls, personal protective equipment, and guidance on safe handling of nanomaterials.
- To enhance global workplace health and safety through national and international collaborations. This involves growing existing partnerships and developing new ones to identify research needs, approaches, and results to ensure worker health and safety.
- Within the context of this program, the potential for disease and injury evaluated by NIOSH goes beyond just consideration of the respiratory tract. However, because respiratory issues are highly integrated within the program as a whole, this section provides an assessment of the entire program and is not limited to respiratory outcomes.

Planning and Production Inputs

To help to achieve its goals, NIOSH has developed a strategic plan to guide the program. NIOSH is also working to coordinate its goals with other research groups, government agencies, and industries. To coordinate nanotechnology research within the Institute, the virtual Nanotechnology Research Center (NTRC) was established in 2004. The NTRC and its steering committee is composed of NIOSH scientists from various disciplines. The role of this committee is to develop and guide the Institute's plan for health research related to nanomaterials in the workplace as well as to guide research on the application and utilization of nanomaterials in occupational safety and health. The project activities managed and coordinated by the NTRC contribute to several sector, cross-sector, and coordinated emphasis areas, including respiratory disease, manufacturing, exposure assessment, personal protective technology, and engineering controls. In keeping with the NORA approach, the nanotechnology research program is managed across the Institute in a matrix fashion through NTRC.

Various members of the nanotechnology research program meet annually to update strategic planning and to review the critical occupational safety and health issues arising from nanotechnology. The latter is a list developed by the NTRC to identify critical issues that may arise from nanotechnology in various areas, including exposure, toxicity, measurement, and control.

NIOSH is a member of the Nanoscale Science, Engineering and Technology Subcommittee (NSET) of the National Science and Technology Council (NSTC) and participates in the National Nanotechnology Initiative (NNI) strategic and budget-planning processes. Through the NSET Nanotechnology Environmental and Health Implications Working Group, NIOSH coordinates research related to the occupational health and safety of nanotechnology with other federal agencies.

Activity, Outputs, and Outcomes

As part of its nanomaterial research effort, in 2004, NIOSH funded the Nanotechnology Safety and Health Research Program, which consisted initially of six research projects aimed at issues of exposure measurement, nanoparticle characterization, and biological effects of exposure in the pulmonary and cardiovascular systems. An additional four projects, aimed at issues related to exposure surveillance, exposure control, risk assessment, and risk dissemination, were added in 2005, and it is anticipated that a further increase in the research portfolio will be forthcoming to address issues of safe handling of nanomaterials, exposure mitigation, and toxicity to workers that may go beyond the respiratory tract as a target organ.

Since this review began, an NTRC progress report for the years 2004-2006 has been released (NIOSH 2007e). The report lists 10 topic research areas that are the core of the NTRC research program to address occupational safety and health issues: toxicity and internal dose, risk assessment, epidemiology and surveillance, engineering controls and personal protective equipment, measurement methods, exposure assessment, fire and explosion safety, recommendations and guidance, communication and education, and applications.

The NIOSH research program related to nanomaterials consists of intramural and extramural activities. Extramural funding for nanomaterial research began in FY2000, and these projects accounted for 3.1% of total extramural funding from that time to FY2006. Funding for extramural projects in FY2007 is lower than that in FY2006, at about $700,000. On the other hand, intramural funding has been increasing rapidly since 2003, and FY2007 funding is $4.6 million.

Activities: Intramural Research Program

The intramural program is concerned with determining nanomaterial exposure concentrations in occupational environments and potential adverse health outcomes from exposure as well as assessing appropriate engineering controls. These activities are reviewed and prioritized by the NTRC Steering Committee and are funded through various sources, including the NORA program, NTRC projects, and supplemental NORA funding. Several projects are listed in the "Strategic Plan for NIOSH Nanotechnology Research" (NIOSH 2005b) and in the recently released progress report (NIOSH 2007e). Funded projects include those on generation and characterization of occupationally relevant airborne nanoparticles, pulmonary toxicity of carbon nanotube particles, the role of carbon nanotubes in cardiopulmonary inflammation and COPD-related diseases, particle surface area as a dose metric, ultrafine aerosols from diesel-powered equipment, nanotechnology safety and health research coordination, nanoparticle dosimetry and risk assessment, nanoparticles in the workplace, web-based nanoinformation library implementation, an ultrafine particle intervention study in automotive production plants, and the filter efficiency of typical respirator filters for nanoscale particles. Given the diversity of types of nanoparticles that may be encountered in occupational settings, it is likely that there will be quantitative and qualitative differences in the manner by which they may affect human health. Furthermore, a wide range of organizations are conducting research on the toxicity of various nanoparticles. Thus, it will be critical for NIOSH to prioritize the areas of research that are most important for its specific consideration.

Activities: Extramural Research Program

Extramural activities are aimed at supporting the strategic plan through research, education, and training. Some of these activities involve collaborating with other agencies, including the National Center for Environmental Research of the EPA, the National Science Foundation, and the National Institute of Environmental Health Sciences of the National Institutes of Health (NIH). As of FY2006, the extramural research project had funded seven projects. Some are joint solicitations from NIOSH and other funding agencies. The proposals are treated as NIH R01 or small business innovation research (SBIR) applications and are reviewed by the NIOSH study section. Research (R01) studies include assessment methods for nanoparticles in the workplace, monitoring and characterizing airborne carbon nanotube particles, lung oxidative stress/inflammation by carbon nanotubes, and the role of surface chemistry in the toxicologic properties of manufactured nanoparticles. Two SBIR studies were funded on new nanostructured sensor arrays for hydride detection and using nanoparticles in lightweight permeable textiles to improve the ability of protective garments to protect against toxic gases. There are also a number of cross-cutting projects that have some utility for evaluating risks from exposure to nanosized particles, although they may not be specifically engineered materials.

Other Activities

A number of activities serve to disseminate information about nanotechnology. They have often been produced in collaboration with NIOSH partners in nanotechnology health and safety research.

- NNI: NIOSH is a member of the NSET of NSTC and participates in the NNI strategic and budget-planning processes. Through the NSET Nanotechnology Environmental and Health Implications Working Group, NIOSH coordinates research related to the occupational health and safety of nanotechnology with other federal agencies.
- NIOSH cosponsored the International Conference on Nanotechnology Occupational and Environmental Health and Safety: Research to Practice in December 2006. This conference addressed the impact of nanotechnology on occupational and environmental health and safety from two perspectives: the promotion and protection of individual heath and safety along the life cycle of nanobased products, and the use of emerging technology in preventing, detecting, and treating occupational and environmental diseases related to nanomaterials.
- NIOSH cosponsored an international symposium, Nano-Toxicology: Biomedical Aspects, in January and February 2006.
- NIOSH cosponsored the Second International Symposium on Nanotechnology and Occupational Health in October 2005.
- NIOSH cosponsored the First International Symposium on Nanotechnology and Occupational Health in October 2004.
- NIOSH published the document "Approaches to Safe Nanotechnology" in 2005 with an update in 2006 (NIOSH 2006f) as an informational exchange to discuss health and safety concerns from exposure to nanomaterials.

Outputs

Safety Document

NIOSH has published a highly requested document on "Approaches to Safe Nanotechnology: An Information Exchange with NIOSH" (NIOSH 2006f), which provides practical advice on the safe handling of nanomaterials by anyone working with the substances. NIOSH has a website to answer frequently asked questions.

Research Publications

NIOSH is just getting started in publishing in this emerging, fast-moving field. Three peer-reviewed articles were mentioned in the evidence package (Maynard and Kuempel 2005; Shvedova et al. 2005; Oberdorster et al. 2005). There were also six abstracts and six proceedings documents relating to manufactured nanoparticles. The agency indicated a much higher number of publications, but they appear to be based on work with diesel particles, and not specifically with manufactured nanoparticles.

The recently released progress report of the NTRC indicates that the number of peer-reviewed publications has increased, with more abstracts and proceedings documents.

Sponsored Conferences

NIOSH has sponsored or cosponsored at least five nanotechnology conferences at the national and international levels.

Invited Talks

NIOSH staff have given about 50 invited talks on nanotechnology at national and international conferences.

Outcomes

This is a fairly new program for NIOSH so outcomes are limited. On an intermediate outcome level, an ISO Technical Committee on Nanotechnology that was established in 2005 has the United States as the leader in the committee's Working Group on Health, Safety and the Environment.

Assessment of Relevance

The nanomaterial research program, as outlined in the "Strategic Plan for NIOSH Nanotechnology Research-Filling the Knowledge Gaps" (NIOSH 2005b), is a highly relevant component of their overall research efforts. The research addresses an important and pressing need to determine the potential toxicity of and methods to monitor and control a newly discovered, highly useful material at the onset of its extensive use in industry. NIOSH has taken the initiative to develop a strategic plan to guide efforts and to allow more effective collaborations with other agencies both nationally and internationally. The focus of the NIOSH nanotechnology program is engineered nanoparticles, but the NTRC supports work within the Institute that can contribute to a better understanding of the behavior, measurement, toxicity, and control of other types of ultrafine particles such as those generated from occupational activities such as welding, diesel engines, and fires. Thus, the work in the nanotechnology program has implications for utility in other aspects of NIOSH.

Assessment of Impacts

This program is dealing with materials having a potential health effect, so measures of any reduction in health effects are not yet determinable. The expected impact is to prevent future health effects that might occur if this research were not done. The committee agrees that NIOSH is very proactive in this area of occupational health, helping to anticipate the need to protect occupationally exposed individuals. This proactivity is manifested both in the development of their research program on nanomaterials as well as in significant outreach as exemplified by publication of the "Approaches to Safe Nanotechnology" document (NIOSH 2006f). Successful development and implementation of this program will provide NIOSH with the opportunity to develop a paradigm to prevent occupationally related illness due to exposure to nanomaterials before their widespread use in industry. This provides NIOSH with a unique opportunity to prevent disease before it occurs rather than to control it after workers develop occupationally related pathology.

Intermediate Impacts

The NIOSH program should ultimately have an impact on setting federal regulations as well as voluntary and professional standards for handling nanomaterials. NIOSH is also having an impact on the education and training for monitoring and controlling nanomaterials in occupational settings.

A major contribution of NIOSH to the latter was the timely publication of the document, "Approaches to Safe Nanotechnology: An Information Exchange with

NIOSH" (NIOSH 2006f), a valuable guide for anyone working with nanomaterials. The leadership of NIOSH in developing this safety manual strongly influenced the appointment of the United States as chair of the Working Group on Safety and Health Standards of the International Standards Organization Technical Committee on Nanotechnologies. NIOSH staff also have helped the EPA to develop a pilot program on the toxicity of nanomaterials and work with the National Institute of Standards and Technology to set standards for nanomaterials.

NIOSH is a leader in providing forums for discussion of the potential health problems associated with the use of nanomaterials and has set up several conferences on the topic. The conference proceedings have been published and the impact of these conferences has been to stimulate further research in the field. NIOSH has interacted well with various stakeholders, including other government agencies and industry, to explore problems associated with nanotechnology.

The output of the NIOSH basic research program on the toxicity of nanoparticles in biological systems is less impressive in that only a few open literature publications have been completed. This probably reflects the newness of the field and the fact that NIOSH, with limited funds, has placed higher priority in areas where NIOSH is particularly well suited to have an impact—that is, in continuing successful research into methods to monitor exposures to nanomaterials and to develop appropriate engineering controls to prevent such exposures. The committee thinks this is appropriate because many other agencies, particularly NIH, are conducting basic research on the health effects of engineered nanomaterials. Nevertheless, the committee encourages continued research efforts on the toxicity of nanomaterials in biological systems as a part of the RDRP program (see emerging issues section). One listed goal is to explore how various applications of nanomaterials might improve workplace safety. This is a limited program confined to a few SBIR extramural projects. The committee agrees that this area should not have a high priority for the RDRP.

End Impacts

It is not possible at this time to assess end outcomes since nanotechnology is a fairly new field and occupational disease has not yet been attributed to exposure to nanomaterials. However, the use of manufactured nanomaterials is expected to grow, and the available database from toxicology studies suggests that nanoparticles may be a potential occupational health hazard. Since there is a potential impact on thousands of workers, it is admirable that NIOSH has taken the lead before development of disease in the workplace; by continuing a leadership role, NIOSH has had and will have a major influence on policy setting.

OVERALL ASSESSMENT OF THE RDRP RELEVANCE AND IMPACTS

A central part of the charge to the committee is to provide a quantitative assessment of the overall relevance and impacts of the RDRP's activities. Using scoring criteria the National Research Council framework committee developed, the committee was tasked with rating the relevance of the RDRP's activities for improving occupational health on a scale of 1 (lowest) to 5 (highest) and rating the impacts of the program's research for reducing work-related hazardous exposures and illnesses on a similar scale. Box 2-1 shows the scoring criteria for relevance and impacts, which was also shown in Chapter 1.

To develop scores for the program as a whole, the committee considered its assessment of the relevance and impacts of NIOSH activities directed at the individual program subgoals described previously. It then weighted these qualitatively to arrive at an overall program assessment. The framework committee recognized the substantial differences among the types of research programs that will be reviewed by the various evaluation committees. It thus declined to provide an explicit instruction on how an evaluation committee should implement the scoring system and weigh the various programs and instead called for individual committees to use their expert judgment to develop its scores.

Relevance Score

As noted earlier, the RDRP is a large program, using nearly $30 million in 2005 in pursuit of its program goals and subgoals. Not surprisingly, the committee found variability in the degree to which individual activities might be relevant to the overall program goals of the RDRP. For example, although the individual activities related to indoor air quality likely contribute to better overall understanding of this issue, their relationship to occupational asthma is not necessarily straightforward. However, there are clearly a large number of programs that are directly relevant to the individual goals. It also is important to note how some activities, especially those that occur outside the Division of Respiratory Disease Studies, may have other NIOSH goals as their primary motivators. For example, RDRP-related activities that occur at the Pittsburgh Research Laboratory must also be relevant to NIOSH's goal of improving mine safety and health.

The committee has assigned a score of 5 in its rating of relevance. This reflects the judgment of the committee that the activities related to most of the subgoals are in the highest-priority subject areas and highly relevant to improvements in workplace protection and that the RDRP is engaged in transfer activities at a significant level. This is particularly true for its activities related to interstitial lung disease as well as many parts of the activities related to airways and infectious

BOX 2-1
Scoring Criteria for NIOSH Program Reviews from Framework Document

Rating of Impact

5 = Research program has made a major contribution to worker health and safety on the basis of end outcomes or well-accepted intermediate outcomes.
4 = Research program has made a moderate contribution on the basis of end outcomes or well-accepted intermediate outcomes; research program generated important new knowledge and is engaged in transfer activities, but well-accepted intermediate outcomes or end outcomes have not been documented.
3 = Research program activities or outputs are going on and are likely to produce improvements in worker health and safety (with explanation of why not rated higher).
2 = Research program activities or outputs are going on and may result in new knowledge or technology, but only limited application is expected.
1 = Research activities and outputs are NOT likely to have any application.
NA = Impact cannot be assessed; program not mature enough.

Rating of Relevance

5 = Research is in highest-priority subject areas and highly relevant to improvements in workplace protection; research results in, and NIOSH is engaged in, transfer activities at a significant level (highest rating).
4 = Research is in high-priority subject area and adequately connected to improvements in workplace protection; research results in, and NIOSH is engaged in, transfer activities.
3 = Research focuses on lesser priorities and is loosely or only indirectly connected to workplace protection; NIOSH is not significantly involved in transfer activities.
2 = Research program is not well integrated or well focused on priorities and is not clearly connected to workplace protection and is inadequately connected to transfer activities.
1 = Research in the research program is an ad hoc collection of projects, is not integrated into a program, and is not likely to improve workplace safety or health.

diseases and malignancies. The committee also noted that the activities related to nanotechnology were highly relevant, even though this emerging area has yet to see any impacts related to intermediate or end outcomes. Activities related to parts of some subprograms, including some of the activities related to malignancies and infectious diseases, do not reach this highest level of relevance as reflected in the assessment of the subprograms earlier in this chapter. But those NIOSH activities were still in important research areas with some connection to improvements in workplace protection.

Impact Score

There is variability in the impacts of RDRP activities on end outcomes or well-accepted intermediate outcomes. Some activities may have large and well-documented impacts, whereas others are smaller and less easily discernable. Again, given the size of the RDRP and the notion that some elements of the RDRP have objectives besides respiratory diseases, the outcome is not surprising. For example, the committee notes that activities related to the development of diagnostic tools for early detection of lung cancer, while relevant to lung cancer in general, have no effect on the program goal associated with work-related respiratory malignancies. The committee recognized that, in terms of assessing measurable impacts, any programmatic efforts of only 10 years' duration could not be expected to be reflected in changing incidence data for respiratory tract diseases with long latency, most notably respiratory tract malignancies. Thus the absence of data for indicating such impacts was not weighted as a "negative" finding. Where appropriate, however, exposure or other risk factor reductions for disease processes with long latencies were considered. Further, there is no way for NIOSH to quantify specifically the impacts that work on personal respirators and environmental controls have had on reductions in the occurrence of TB and other respiratory infections. However, the activities of the RDRP have clearly played major roles in reducing the occurrence of and mortality from CWP. It also played a major role in reducing the prevalence of latex sensitization as a result of the intervention effort that began with the 1996 NIOSH Alert.

The committee has assigned a score of 4 in its rating of impact, reflecting the committee's judgment that most of the subprograms within the RDRP have made major contributions to worker health and safety on the basis of end and well-accepted intermediate outcomes. It represents the consensus of the committee on the degree to which the overall program, which is still in its infancy, meets the goals and has had the impacts set out by NIOSH. After much deliberation on how to weigh the assessments of different subprograms, the committee assigned a score of 4 for the program as a whole. Had the committee been given the option of providing non-integer scores, the score for program impact would have been between 4 and 5, based on consensus that the program was clearly better than that called for in a score of 4 but not in sum what the committee would rate a 5.

3

Emerging Research and Research Issues

The goal of this chapter is to assess progress in targeting new research relevant to future improvements in workplace protection from occupational respiratory diseases and identifying significant emerging research areas important to the mission of the National Institute for Occupational Safety and Health (NIOSH).

The NIOSH evidence package (NIOSH 2006a, Chapter 13) provides a "vision for the future" statement that summarizes past accomplishments and their link with future plans in terms of their five strategic goals. This chapter serves as a useful framework within which the evaluation committee will present its findings.

We first present a brief summary of the Respiratory Diseases Research Program (RDRP) presentation of important plans for its five goals, each of which is followed by the evaluation committee's assessment; then we discuss cross-cutting issues the committee identified as important considerations for future NIOSH-RDRP research work.

STRATEGIC GOAL 1: PREVENT AND REDUCE WORK-RELATED AIRWAY DISEASES

Summary

The RDRP gives very high priority to this area, with a focus on improved characterization, exposure assessment, and elucidation of the mechanisms of action of low-molecular-weight agents and mold allergens. Within this goal are a plan to implement and monitor a model screening and surveillance program for work-related asthma (WRA) in the isocyanate industry, an evaluation of the effectiveness of recommendations made in response to health hazard evaluations (HHEs), and continued development of better approaches to facilitate the use of ambulatory spirometry for assessing WRA. With regard to work-related chronic obstructive pulmonary disease (COPD), the RDRP will continue to evaluate the relation between COPD prevalence and occupation and pursue aggressive surveillance of work-related bronchiolitis obliterans. Much of this goal is a direct extension of material presented in the evidence package. The focus on mold allergen stems from the NIOSH-RDRP response to the high frequency of HHE requests triggered by asthma and asthma-like symptoms.

Committee Comments

Exposure and disease surveillance are critical to the objective of detecting and reducing work-related airway diseases as well as other major program goals of the RDRP. WRA is a difficult condition to identify and the RDRP should explore ways to identify new agents, outbreaks, or "hot spots" of WRA caused by known agents and trends in WRA. Furthermore, many WRA agents are present in small businesses that current surveillance methods do not identify or track effectively. Expanding the SENSOR program to include additional geographic areas and, thus, an even broader occupational mix would provide the much needed opportunity to track trends and to highlight possible geographic variation. One strength of the SENSOR program is that it does have elements that go beyond a purely passive (that is wholly voluntary) scheme.[1] Active elements, such as mandatory physician reporting, warrant additional follow-up from the RDRP to evaluate their possible extrapolation to other geographic areas and other respiratory diseases. Without SENSOR data, the appropriate targeting of limited resources for exposure and medical monitoring and interventions to prevent and control disease is difficult, if not impossible. NIOSH should develop strategies to work with state health departments to develop methods for WRA surveillance.

The requirement for improved surveillance in work-related COPD parallels the need with WRA. RDRP efforts to seek new methods for conducting work-related COPD surveillance are of great importance. Furthermore, the RDRP should

[1]The California SENSOR program, in particular, has utilized legally required physician reporting of occupational disease (the Doctor's First Report system) for case finding. In addition, a NIOSH-led effort exploited the multiple routes of SENSOR case detection in the Michigan program to carry out a formal "capture-recapture" analysis of case incidence, once again highlighting that this scheme transcends the strict definition of a passive surveillance program (Henneberger et al, 1999). Also in Michigan, NIOSH participated in a similar surveillance related to silicosis (Rosenman et al 2003).

explore ways to identify new agents or occurrences of work-related COPD. Efforts to ensure spirometry in the next round of the National Health and Nutrition Examination Survey (NHANES) are encouraged, as the lack of lung function would be a major setback for COPD surveillance at the national level. The ability to detect work-related COPD requires including new questions on occupational exposures as well as spirometry in the NHANES survey. An important area of work-related COPD research is the development of improved methods for analyzing pulmonary function data from longitudinal studies.

Finally the committee notes that, while the RDRP has made critically important contributions to preventing exposure to latex and considering isocyanate as a cause of WRA, the program has not used these successes to develop a more systematic approach to the classes of high- and low-molecular-weight sensitizers. The committee considers this to be an area that should be articulated more coherently. Such a conceptualization is well suited to the matrix organization under the second National Occupational Research Agenda (NORA2).

STRATEGIC GOAL 2: PREVENT AND REDUCE WORK-RELATED INTERSTITIAL LUNG DISEASES

Summary

Because of the recent identification of clusters of miners with progressive massive fibrosis, the RDRP plans to work on further technological improvements to reduce exposure, ensure the availability and use of personal dust monitors, and investigate the nature and causes of hot spots of coal workers' pneumoconiosis (CWP). A move to digital radiography for application of the International Labour Office (ILO) classification of CWP will not occur until there is further assessment, validation, and acceptance of digitized films for use in the ILO classification. Work will continue on the development of molecular biomarkers for silica exposure for early detection. The new digital radiography programs are seen as being relevant to more intense medical screening and surveillance of silicosis. Collaboration with Brush Wellman will continue with the collection of additional data allowing for reanalysis of epidemiologic findings to examine dose-response relations in light of the estimated dose of dissolved beryllium and dermal exposure. This effort is motivated by analyses that suggest mass-based exposure is not optimal for predicting disease risk. The RDRP will create transgenic mice to examine the impact of dose, form, and route and associated pathology and pathophysiology. These mice will be made available to other investigators.

Committee Comments

The committee's review of the RDRP's research on interstitial lung disease generally documents a sustained, thoughtful, and effective research program that has had a marked impact on respiratory exposures and the incidence of interstitial diseases in the numerous workplaces where these diseases occur. Much of the research and surveillance activities dealing with CWP was mandated by the Federal Coalmine Health and Safety Act of 1969. These programs, which were established in the 1970s, led to a marked reduction in miners' exposure to respirable coalmine dust and until recently have decreased the annual prevalence of pneumoconiosis.

Measures have been taken to improve the participation of miners in mandated surveillance efforts. Surveillance data have recently identified a cluster of rapidly progressing CWP in Southern West Virginia and Eastern Kentucky. Targeted hot spot surveillance has been undertaken to further elucidate the reasons for rapid progression of CWP among miners with relatively short exposures. This experience should lead to a full reexamination of the organization and efficacy of the CWP surveillance program that includes the interaction between NIOSH and the Mine Safety and Health Administration (MSHA), with additional focus on the adequacy of exposure assessment and compliance determination.

Additional factors that need better understanding include the possible role of coal rank, silica level, and duration of exposure in the mines producing these rapidly progressive cases of CWP. These issues need to be assessed further epidemiologically and even experimentally. Assessment of digital radiography in CWP surveillance is also an important continuing research priority. Other emerging issues more directly tied to CWP include exposure levels in light of changing work patterns underground, especially longer work shifts; the impact of deeper underground mining conditions on dust exposures and lung health; and the respiratory risks (beyond physical trauma) that may result from "retreat mining."

The RDRP's work related to pulmonary toxicity from silica, which includes traditional and innovative approaches to silica control and exposure assessment, and work on the basic mechanisms of oxidant injury resulting in interstitial lung disease have been important contributions by RDRP scientists. More work remains to be done with regard to elucidating the toxic and genetic mechanisms of silicosis and the issue of lung cancer arising from exposure to silica. A strong and continuing emphasis is needed on control technology and on efforts to reduce silica exposures in multiple mining and nonmining workplaces.

NIOSH also has played an important role in contributing to our understanding of the epidemiology of exposure to commercial asbestos products. This work led to a revised NIOSH recommended exposure limit in 1976 and recommendations with regard to asbestos substitute materials. Despite the well-understood epidemiology of asbestos exposure and significant progress in the control of commercial asbestos exposures, there remains a very important need for RDRP scientists to be engaged in understanding the differential risk from exposure to specific subtypes of asbestos as well as methods for surveillance of asbestosis and cancer related to asbestos exposures as these conditions continue to be reported in exposed workers.

NIOSH appropriately has concentrated its efforts on asbestos exposures in asbestos-contaminated vermiculite that was widely used in housing and commercial building insulation and for many commercial products. This RDRP research has led directly to controlling these exposures through regulation and ultimately closing the mine that provided this contaminated vermiculite for commercial use.

NIOSH has contributed to our understanding of respiratory disease among nylon flock workers. Description of this interstitial lung disease arose from collaborative research with university investigators and appropriately included epidemiology, industrial hygiene, control technology, and experimental studies leading to a well-documented newly described interstitial lung disease. RDRP investigators have worked effectively with the flock industry to implement controls. While surveillance of this industry apparently shows no new cases, subclinical cases have been found and continued surveillance is needed.

NIOSH has also contributed to our understanding of possible pulmonary consequences arising from refractory ceramic fibers (RCFs). Monitoring of exposure and control technology research has led to a product stewardship program, which RDRP scientists continue to monitor. This voluntary program with the Occupational Safety and Health Administration (OSHA) that arose in conjunction with the NIOSH RCF criteria document is an alternative to the traditional regulatory approach and is innovative, but it needs continuing evaluation of its efficacy in documenting and controlling risk. RDRP research on fiber characterization and toxicity also is appropriate and has aided regulatory and other federal agencies and international bodies in their assessment of fiber ranking with regard to carcinogenicity in humans. These efforts should continue with attention to low-level exposures and unresolved issues such as cleavage fragments.

Overall, the RDRP's documentation of health effects from respirable fibers demonstrates the importance of surveillance to detect effects and research to understand the processes resulting in exposures. The experience with nylon flock shows the potential for the emergence of previously unknown fiber-related diseases and, ultimately, the importance of evaluating the potential negative consequences of new fibers or new uses of fibers in occupational settings. NIOSH is encouraged to continue and expand its surveillance activities to evaluate the potential effects of synthetic fibers such as nylon flock and RCF.

NIOSH has made substantial contributions to our understanding of sensitization and the risk of chronic beryllium disease from industrial exposures. Industrial exposure to beryllium continues to be widespread and the population at risk is relatively large. NIOSH epidemiologic work has documented that the OSHA permissible exposure limit (PEL) does not prevent sensitization or the onset of chronic beryllium disease. This highlights the need for OSHA to recommend a new PEL based on RDRP research that will satisfactorily protect workers. NIOSH's collaboration with Brush Wellman appears to provide incentives for the industry and high-quality research for NIOSH and the research and regulatory communities.

NIOSH has evaluated the importance of the interaction of occupational exposures and genetic factors by looking specifically at the influence of genetic variation on chronic beryllium disease. The research has been productive and it is tempting to consider this relationship as a model for understanding gene-environment interactions in disease causation. However, the committee recognizes that research on both genetic and epigenetic relationships to the environment is likely to prove much more complex than this one example (Kelsey 2007; Schulte 2007; Vainio 2007). Furthermore, how research on these interactions (whether on beryllium or other toxicants) can be applied to protect exposed workers or assist in protecting susceptible subpopulations of workers remains unclear (Palmer et al. 2004), although the research will likely lead to better understanding of the molecular mechanisms driving workplace-related disease. The current program articulated by RDRP scientists is appropriate but should not lose sight of the need for an OSHA-recommended standard and a mandated comprehensive beryllium control program for all workplaces where this exposure occurs.

STRATEGIC GOAL 3: PREVENT AND REDUCE WORK-RELATED INFECTIOUS RESPIRATORY DISEASES

Summary

Future efforts in this area largely are a continuation of the work being carried out by the RDRP. Engineering controls, improved personal respirator protection, and preparation for possible pandemic and bioterrorism events remain the focus. Surprisingly, the efforts directed toward understanding the effects and mechanisms of exposures such as diesel exhaust particles on pulmonary susceptibility to infection are scarcely mentioned, despite the considerable emphasis placed on them in the evidence package.

Committee Comments

One important area that is missing from this strategic goal relates to surveillance activities. While surveillance for the occurrence of infectious diseases is acknowledged as an important component of preventing and reducing infectious diseases (particularly for tuberculosis), the RDRP has not proposed a plan to address the lack of surveillance for infectious diseases in occupational settings. The SENSOR program, described in Chapter 2, attempted to address some of these surveillance deficits. However, the program did not appear to have a specific infectious disease component. Moreover, the SENSOR program is being downsized. The RDRP Infectious Diseases Program does not appear to have the resources to develop alternatives to offset the loss of this program and will have to depend on surveillance data from programs not directly under its control.[2]

Another component of effective surveillance is the rapid detection and identification of disease causing agents when they are present. Although NIOSH has made some efforts in this regard, such as developing sampling procedures and technologies for anthrax, other federal groups, such as the Department of Homeland Security (DHS), support and conduct extensive research to develop technologies for the rapid detection of bioterrorism agents (see, for example, the DHS technology forecast [HSRC 2006] and its "lab-in-a-box" technology that could be adapted to the RDRP mission; the technology is briefly described in USDHS [2007]). It is important that NIOSH participate in initiatives within the federal complex that keep it informed of detection methods under development within other federal agencies. As these technologies become available, it will be necessary to extend their use to protect and reduce infectious diseases in workers.

Three strategies—preventing infection with respirator controls, understanding the mechanisms that underlie susceptibility, and conducting robust surveillance for disease outbreaks—represent NIOSH's primary tools to prevent and reduce known and unknown (emerging) respiratory infectious diseases. Support for respirator control and understanding susceptibility is encouraged to maximize worker protection from emerging diseases, and these areas should remain a high priority, while surveillance activities for infectious diseases in occupational settings need to be increased.

---

[2]NIOSH explained that there are several sources of data (e.g., data collected by OSHA or components of the Centers for Disease Control and Prevention) that can be used in general surveillance for occupational infectious diseases, but these data-collection activities are not managed under a single programmatic umbrella (NIOSH 2006b). Several relevant data sources are described in Appendix A of the NIOSH *Worker Health Chartbook* (NIOSH 2004); however, the availability and content of these resources are limited (Sepkowitz and Eisenberg 2005).

STRATEGIC GOAL 4: PREVENT AND REDUCE WORK-RELATED RESPIRATORY MALIGNANCIES

Summary

Work to reduce exposure to silica will continue. In addition, diesel exhaust and mineral fibers are targeted for future focus. Cooperation with the National Cancer Institute (NCI) epidemiologic study of diesel-exhaust-related mortality is identified as a source of exposure-response data with respect to lung cancer. Electron microscopy will be used to reassess exposure-response relations that account for asbestos fibers that could not be observed with light-microscopy-based methods to assess exposure; data from previously studied cohorts will be used to provide new exposure-response estimates for cancer development. Laboratory studies will be conducted to evaluate the impact of fiber characteristics on carcinogenesis. Such studies are considered particularly important for nonasbestos fibers.

Committee Comments

The evaluation committee finds that the RDRP has provided evidence for a solid research program that has had a direct impact on the control of occupational cancers. This work has been of direct use to OSHA and MSHA in their standard-setting processes, particularly for hexavalent chromium and diesel particulate, as well as to the International Agency for Research on Cancer in developing its assessment of cancer risk from occupational exposures. The work on silica also is considered very strong.

Ongoing research continues to address challenging problems related to the risk of occupational lung cancer, and the RDRP has been effective in engaging stakeholders from industry and the workforce. In particular, the RDRP's efforts have added substantially to dissemination of research findings from cohort studies to individual workers. Collaboration with industry is often the only way to understand workplace exposures that lead to cancer. The committee supports ongoing and encourages future collaborations in studies of occupational lung cancers, particularly in the face of emerging evidence of the link between exposure and outcome.

For some time, the epidemiologic literature from cohort studies of textile workers has provided evidence that workers exposed to cotton dust have lower than expected rates of lung cancer (Henderson and Enterline 1973; Merchant and Ortmeyer 1981; Hodgson and Jones 1990; Wernli et al. 2006). In the 1980s, endotoxin was suggested as the source of reduced risk (Enterline et al. 1985). Recently, interest in this association resurfaced when deficits in lung cancer were observed in a large cohort study of cotton workers in China and the deficit was found to be dose-related based on quantitative estimates of historic exposures to endotoxin (Wernli et al. 2003; Astrakianakis et al. 2007). These findings, although intriguing, suggest that further work is necessary to understand the potential mechanisms of the cancer-preventive action of endotoxins, if they exist (Boffetta 2007). Exposure to endotoxin is sufficiently common in a wide range of industrial settings (ranging from agriculture to metal machining) to warrant consideration by the RDRP of what it might be able to contribute to understanding the nature and importance of the interaction of endotoxin and cancer. NIOSH's long-term interest in the effects of cotton dust and the measurement of endotoxin suggests this as a potential fruitful area for research.

As described in Chapter 2, NIOSH and the NCI have undertaken a major epidemiologic study of diesel exhaust and mortality to provide essential epidemiologic information on the quantitative relationships between diesel exhaust and cancer, particularly lung cancer. Progress in the study, however, has been affected by intensive scrutiny and legal action by the industry and congressional intervention (Monforton 2006). The number of workers exposed to diesel exhaust and the potential for this study to provide significantly improved information about the nature and extent of the risk requires that the study be completed with deliberate speed. The design and conduct of the NIOSH-NCI diesel study are meeting high levels of scientific quality. It is important that the peer review process proceed unhindered to allow data to be used by the scientific and regulatory community.

STRATEGIC GOAL 5: PREVENT RESPIRATORY AND OTHER DISEASES POTENTIALLY RESULTING FROM OCCUPATIONAL EXPOSURES TO NANOMATERIALS

Summary

The RDRP plans to focus its attention on the relative toxicities of different nanomaterials in biological systems. The RDRP plans to develop and validate methods of exposure assessment and recommendations for appropriate medical monitoring and surveillance and indicates that it will consider recommendations for engineering controls but does not articulate a specific research program.

Committee Comments

The growing recognition of the usefulness of nanomaterials in various industrial applications and their increased prevalence has created an urgent need to study (1) the potential health effects of exposures to nanoparticles, and (2) methods to control exposures to nanoparticles during manufacturing processes. The RDRP has taken a lead at the national and international levels to address these questions. Such studies require a multidisciplinary approach that includes expertise in the physical and biological sciences to develop methods to characterize the particles physically and chemically, methods to develop adequate protective gear to prevent exposures, methods to maintain the characteristics of the particles during inhalation exposures of test animals, and methods to detect biological responses to exposure to the particles. NIOSH is one of only a few facilities with such a broad multidisciplinary expertise in one agency.

The RDRP proposes to conduct a dose-response inhalation toxicology study in animals with engineered nanoparticles. This should be a valuable contribution and provide the information needed to develop a quantitative risk assessment model. The RDRP also plans to continue to conduct exposure assessments and engineering control evaluations in nanomaterial production facilities. These areas of work should provide valuable information. The committee finds that these are commendable and appropriate future plans that appropriately seek to assess and address the potential emerging effects of nanoparticles.

CROSS-CUTTING ISSUES: CONTINUING AND EMERGING ISSUES CUTTING ACROSS MULTIPLE PROGRAM GOALS

During its deliberations and review of the materials present in the evidence package and in several in-person presentations, the committee identified several continuing and emerging issues that cut across all program goals. These issues include surveillance, exposure assessment, emergency response, respirator policy, and the RDRP's resource allocation. In some instances, the committee thought that the RDRP's attention to these issues was not articulated very clearly or in ways that could motivate and strengthen current and future research programs. The committee will provide recommendations with regard to cross-cutting issues in Chapter 4. Here, we provide a brief summary of these issues.

Surveillance

The evidence package and public presentations made by NIOSH-RDRP indicate clearly the importance of surveillance to the research, dissemination, and prevention activities of the RDRP and the limitations with respect to resources for surveillance under which the RDRP is constrained to operate. The committee understands that a lack of financial and personnel resources and not a lack of awareness or expertise are the major causes of inadequate surveillance. While the needed resources are not likely to emerge in the near future, the committee thinks that it is important to highlight the limitation as a continuing and emerging issue. The performance assessment of current and future RDRP activities and the identification of new respiratory disease issues depend on having an effective surveillance program.

Exposure Assessment

Exposure assessment is a core component of occupational respiratory disease research and prevention activities. The field of exposure assessment incorporates several closely related areas that include instrument development, modeling, and the use of exposure biomarkers. The RDRP evidence package includes excellent examples of relevant and high-impact contributions in all these areas as well as specific future goals related to exposure assessment, as noted previously. However, the RDRP does not present these exposure assessment activities as part of an explicit or comprehensive focus on exposure assessment methods, leaving it to the reader to make the extrapolation. Further, while a number of RDRP scientists focused on exposure assessment research in the past, current or future staffing and programmatic needs for exposure assessment activities are not specifically mentioned.

Emergency Response

It is unclear whether the RDRP plans to incorporate research activities related to emergency responses into future efforts. In particular, much could be learned about exposure-response relationships, and ultimately protecting emergency responders and exposed populations, if appropriate resources were devoted to careful modeling of spatial-temporal toxicant concentrations and estimation of personal exposures for longitudinal cohort studies of individuals involved in catastrophic events (e.g., those from the World Trade Center [WTC] or work settings where sudden exposures to high levels of irritants are experienced [see Chapter 2, "Emergency Response," for further discussion]). In follow-up studies, data derived from emergency responses to toxicant exposures should be applicable to models of irritant-induced asthma, fixed airway obstruction, interstitial lung disease, and possibly even malignancies. The RDRP is encouraged to explore research strategies in their emergency response efforts.

Respirator Policy

NIOSH respirator development, testing, and certification efforts continue to play a crucial role in preventing WRA and work-related COPD. This includes protection from tuberculosis and protection from pandemic and avian influenza and microbial agents that could be used in a terrorist attack. No discussion of airway-disease-specific respirator policy was included in the evidence package, although examples of RDRP work on improvements in the fit and availability of personal respirators were documented. A NIOSH report in the peer-reviewed literature that supported the use of respirators in emergency response situations that involve irritant dust exposures (e.g., the WTC disaster) is an example of how the agency can highlight the importance of effective respirator policy in preventing work-related airway disease (Feldman et al. 2004). Given the need for respirators in such situations as well as for their possible use against other agents, the development of policies related to respirator use is a continuing issue. A separate National Research Council committee is reviewing NIOSH's personal protective technology research program; therefore, recommendations on this topic will be addressed by that committee and are not discussed in Chapter 4.

RDRP Resource Allocation

As noted earlier in this report, the formal RDRP is a recent creation organized during the NORA2 process. Given that the NORA2 process emphasized sector-based as opposed to disease-based research, an emerging issue is how research priorities for respiratory diseases that cut across sectors will be treated. Further, although the focus of respiratory disease research remains within NIOSH's Division of Respiratory Studies, the RDRP encompasses many divisions and laboratories across NIOSH. For researchers who are part of the RDRP but outside the Division of Respiratory Studies, an issue will be how well their activities can be coordinated and prioritized within the RDRP. Further, the new RDRP is faced with the need for systems to govern the awarding of extramural grants, contracts, and cooperative agreements and integrating the results of this external research into the intramural program.

4

Recommendations

The previous chapter summarized the evaluation committee's reviews of the components of the National Institute for Occupational Safety and Health (NIOSH) Respiratory Diseases Research Program (RDRP) and discussed emerging issues related to each of the five goals listed by the program in the evidence package. This chapter provides the RDRP with specific recommendations for each of the strategic goals and for issues that cut across the programs. Major recommendations are summarized in Box 4-1. Greater detail is then presented for each goal and cross-cutting issue. For each of them, a brief summary of the goal or issue and the potential impact of relevant activities is presented; then specific recommendations are provided for the RDRP to consider when planning future work. Finally, recommendations on broader programmatic goals are presented.

STRATEGIC GOAL 1: PREVENT AND REDUCE WORK-RELATED AIRWAY DISEASES

Work-Related Airway Diseases

The RDRP has divided work-related airway diseases into two main categories, work-related asthma (WRA) and fixed obstructive airway disease. The RDRP identified four subgoals related to WRA in the evidence package. Because the contribution of occupational exposures to the burden of adult asthma is high, work in pursuit of the four WRA subgoals can have a potentially large impact through improved occupational safety and health among the U.S. workforce. This potential has been demonstrated by previous RDRP WRA activities. The evaluation committee is concerned, however, that the new second National Occupational Research Agenda (NORA2) industrial-sector-based priority-setting approach may lead to a decreased emphasis on needed asthma-focused research.

Recommendation: The RDRP should systematically evaluate whether WRA activities are being weakened under the new NORA2 approach.

Other specific recommendations to enhance the relevance and impact of the RDRP's WRA research activities are organized by the subgoals as delineated by NIOSH.

Subgoal: Preventing and reducing natural rubber latex asthma and allergy among health care workers.
Recommendation: The RDRP's efforts to prevent latex allergy and asthma have been highly successful. RDRP investigators have documented that the prevalence of latex sensitization fell as a result of the intervention effort that began with the 1996 NIOSH Alert. The evaluation committee recommends that the RDRP assess how its successful work on latex and asthma can be extended to other high-molecular-weight sensitizers that cause occupational asthma and occupational rhinitis.

Subgoal: Preventing and reducing WRA in the isocyanate production industry.
Recommendation: Previous RDRP work on diisocyanates directly addressed the most common low-molecular-weight sensitizing cause of asthma in the developed world and led to important knowledge that has been transferred to prevent disease. The committee recommends that the RDRP assess how its research on diisocyanates and asthma can be extended to other low-molecular-weight sensitizers that cause occupational asthma, especially in terms of mechanisms of disease.

Subgoal: Preventing and reducing WRA related to nonindustrial indoor environmental quality.
Recommendation: While the indoor environmental quality work of the RDRP is judged to be relevant to occupational health and safety in the general sense, it is not always clearly related to WRA. The committee recommends that the RDRP reexamine whether its indoor air-quality-related research is sufficiently relevant to work-aggravated asthma. Moreover, the RDRP should reevaluate the relative commitment of resources to indoor air-quality investigations, as the health effects are often not airway in nature (that is, systemic or neurologic complaints).

Box 4-1
Summary of Recommendations

Strategic Goal 1: Prevent and Reduce Work-Related Airway Diseases

Improve detection of work-related asthma, work-related fixed obstructive airway disease, and relevant exposures. The evaluation committee is concerned that the new, second National Occupational Research Agenda industrial-sector-based priority-setting approach may lead to a decreased emphasis on needed disease-focused research related to airway diseases. The RDRP should systematically evaluate whether work-related asthma (WRA) activities are being compromised under the new approach. Because the contribution of occupational exposures to the burden of adult asthma is high, work in pursuit of the four WRA subgoals (preventing and reducing rubber-latex asthma and allergy among health-care workers, preventing and reducing work-related asthma in the isocyanate-production industry, preventing and reducing work-related asthma related to nonindustrial indoor environmental quality, and improving detection of work-related asthma and relevant exposures, described in more detail in Chapter 4), can have a potentially large impact on improved occupational safety and health among the U.S. workforce. In terms of chronic obstructive pulmonary disease (COPD), understanding the contribution of occupational exposures is difficult. To understand this issue, the evaluation committee strongly recommends that, for planning preventive strategies, the RDRP continue to support population-based studies of associations between occupational exposures and COPD to better define groups of workers at greatest risk. In the flavoring industry, the RDRP response to the identification of diacetyl-induced bronchiolitis obliterans has led to surveillance efforts in multiple locations in an effort to detect and prevent disease. The evaluation committee agrees that continued surveillance, prevention of exposures, and mechanistic research to better understand this disease should continue to be a high priority for the RDRP.

Strategic Goal 2: Prevent and Reduce Work-Related Interstitial Lung Diseases

Continue and expand efforts to prevent coal workers' pneumoconiosis (CWP), silicosis, fiber-induced interstitial lung disease, and chronic beryllium disease. The activities related to interstitial lung diseases form a critical core of the RDRP and have provided well-documented improvements in occupational health. It is important that the RDRP continue to expand its activities in these areas so it can build on its earlier successes while responding to new challenges. In particular, the committee recommends giving high priority to research into the increasing incidence of CWP, "hot spots" of rapidly progressive CWP, and the possible role of concentration and duration of exposure, coal rank, and silica level in the rapidly progressive cases of CWP. Other important areas of research recommended by the evaluation committee are experimental studies of silica-induced cytotoxicity and fibrogenesis and the development of control technologies, including silica substitutes, particle surface coatings, and dust reduction measures. In addition, fibers that are asbestiform, such as vermiculite, or of synthetic origin, such as nylon flock and refractory ceramic fiber, require continued study with attention to fiber characteristics, such as cleavage fragments and low-level exposures, respectively. The committee recommends that the RDRP target work in support of a new recommended OSHA standard that would lead to improved controls to reduce airborne and dermal exposure to beryllium in all workplaces where it is used. The committee recommends that the effectiveness of digital radiography in CWP surveillance should be an important continuing research priority, which will extend to all interstitial lung diseases.

Strategic Goal 3: Prevent and Reduce Work-Related Infectious Respiratory Diseases

Continue to support efforts to protect workers from occupational exposures and to define mechanisms that make workers susceptible to respiratory infections. Enhance surveillance for outbreaks of known occupational respiratory infections as well as emerging respiratory infections. Develop an overarching structure for the infectious disease component and coordinate with other federal agencies to adopt technologies for the detection of bioterrorism agents for the protection of workers. The RDRP's efforts on infectious diseases are appropriately concentrated on preventing infection with respirator controls and understanding the mechanisms that underlie susceptibility. More robust surveillance for disease outbreaks is needed in occupational settings. These three approaches represent NIOSH's primary tools to prevent and reduce known and unknown (emerging) respiratory infectious diseases. The committee recommends that the RDRP collect specific occupational TB surveillance data and explore ways to improve TB surveillance. It also should consider dropping its subgoal of preventing outbreaks of histoplasmosis, because no new RDRP research is planned and no resources are available for specific surveillance activities.

Strategic Goal 4: Prevent and Reduce Work-Related Respiratory Malignancies

Develop a comprehensive plan for addressing respiratory malignancies in the workplace while assuring the integration of this plan with NIOSH and other federal agency research program efforts to study malignancies. Refocus research on diagnostic tools to research on biomarkers of exposure or early detection of risk specific to occupational cohorts. The impact of the respiratory malignancies program has been strong with regard to the three specific carcinogenic exposures listed as subgoals by the RDRP: hexavalent chromium, silica, and diesel exhaust. Ongoing research on respiratory malignancies resulting from workplace exposures continues to address challenging problems related to occupational lung cancer risk, and the RDRP has been effective in engaging stakeholders from industry and the workforce. To enhance its efforts, the RDRP needs to develop, in collaboration with other relevant federal agencies (National Cancer Institute, National Institute of Environmental Health Sciences, Food and Drug Administration, Department of Defense, Department of Labor), a coordinated planning process to address occupational cancer risks to provide a

continued

Box 4-1 Continued

comprehensive approach to detection, surveillance, and prevention. The committee recommends that the RDRP ensure that respiratory malignancies are well integrated into an overall program of occupational cancer research, and not arbitrarily separated from those efforts. The RDRP should consider refocusing its research on biomarkers for early detection to biomarkers of exposure or to early detection that addresses the needs of specific workers at high occupational risk of contracting lung cancer. The RDRP should continue efforts to develop and validate exposure methods for diesel particulate matter, especially in the presence of other sources of carbon aerosols, and to validate the methods used to measure diesel particulate matter in coal mines. The RDRP should consider developing long-term follow-up of workers exposed to asbestos and should consider enhancing surveillance for asbestos-related risks.

Strategic Goal 5: Prevent Respiratory and Other Diseases Potentially Resulting from Occupational Exposures to Nanomaterials

NIOSH should continue to play a leading role in informing and guiding national and international efforts to address potential occupational hazards and risks associated with the use of manufactured nanomaterials. The growing recognition of the usefulness of nanomaterials in various industrial applications has created an urgent need to study the potential health effects of exposures to nanoparticles and methods to control exposures to these particles. The RDRP has taken a lead at the national and international levels to address these questions. The RDRP is well-suited to continue to develop exposure assessment methods and technology to monitor effective control of exposures to nanomaterials in work settings. The committee generally supports the RDRP's research efforts on nanomaterial toxicity, exposure, and dose-response as part of a coordinated effort with other federal agencies and with appropriate prioritization for resource allocation to this problem. However, the committee is concerned that available data—especially on human-health effects—might not be sufficient for quantitative risk assessments, and therefore that the RDRP should consider other approaches for dealing with the potential health impacts of these new materials in a precautionary manner.

Cross-Cutting Issues

Systems for Surveillance: *NIOSH should provide appropriate resources for and engage in high-priority occupational disease surveillance.* The United States is practically alone among highly developed countries with regard to its lack of comprehensive surveillance of occupational diseases. The effectiveness of past NIOSH surveillance activities for coal-dust-related diseases, both CWP and COPD, highlight the importance of improved surveillance for other occupational respiratory disorders. NIOSH should engage in development and evaluation of surveillance methods as a high priority for surveillance of occupational respiratory disease, including methods to systematically review and analyze the findings of reports through the health hazard evaluation and technical assistance program and other data-collection approaches.

Exposure Assessment: *Develop a programmatic approach to the development of sampling and analytic methods that include exposure assessment scientists as an integral part of RDRP activities.* Exposure assessment is a core component of occupational respiratory disease research and prevention activities. However, the RDRP does not present an explicit or comprehensive focus on exposure assessment methods. Also, while the RDRP has focused on exposure assessment research in the past, no specific mention of current or future needs for exposure assessment activities is made explicitly.

Emergency Response: *The RDRP is encouraged to explore research strategies in its emergency response efforts. For example, the RDRP should assess how NIOSH-supported research and medical surveillance of World Trade Center disaster emergency responders and recovery workers may or may not be relevant to WRA, work-related fixed obstructive airway disease, interstitial lung disease, and possibly malignancies.* The RDRP has made important contributions to the emergency response to recent disasters, including the World Trade Center and anthrax terror attacks, and hurricanes Katrina and Rita. However, much more could be learned about exposure-response relationships and ultimately about protecting emergency responders by conducting longitudinal cohort studies relating to catastrophic events. Information from emergency responses to toxicant exposures should be applicable to models of irritant-induced asthma, fixed airway obstruction, interstitial lung disease, and possibly even malignancies. The RDRP is encouraged to continue to develop cooperative work with other agencies that have mandates in infection and terrorism.

RDRP Resource Allocation: *The RDRP should prioritize all research proposals under consideration for funding, whether intramural or extramural, according to the RDRP strategic plan, which needs to be updated periodically.* The RDRP has recently been organized to emphasize sector-based as opposed to disease-based research; an emerging issue is how research priorities for respiratory diseases that cut across sectors will be treated, particularly since the RDRP encompasses many divisions and laboratories across NIOSH. The RDRP needs systems to govern the awarding of extramural grants, contracts, and cooperative agreements and integrating the results of this external research into the intramural program. This system should ensure that unnecessary duplication and inappropriate expenditure on low-priority research projects are avoided.

Subgoal: Improving detection of WRA and relevant exposures.
Recommendation: The RDRP effort to improve the detection of WRA is of the highest relevance. While the quality of transfer activities to increase the awareness of WRA has been high, the evaluation committee recommends that greater attention be paid to irritant-induced asthma given its relative importance as demonstrated by SENSOR data. Irritant-induced asthma warrants specific RDRP planning and goal setting that would build on accomplishments already made in this area. In addition, because of the key role of SENSOR data in surveillance for WRA, the RDRP should consider aggressively expanding this program beyond the collaborations that currently exist with four states as well as including "active" elements of surveillance.

Chronic Obstructive Pulmonary Disease

The RDRP identified four subgoals related to fixed obstructive airway diseases. As noted for WRA, the likely contribution of occupational exposures to the burden of chronic obstructive pulmonary disease (COPD) is high, and thus work in pursuit of the four fixed obstructive airway disease subgoals potentially can have a great impact on improved occupational safety and health among the U.S. workforce.

Subgoal: Establish the work-relatedness of COPD.
Recommendation: The previous research of RDRP investigators on the risk of COPD due to exposure to coal dust has been cited by a number of investigators and policy makers in their assessment of the links between dust exposure and COPD, not only for this specific industry but also as a measure of biologic plausibility for COPD related to other exposures. RDRP studies using data from the National Health and Nutrition Examination Survey (NHANES) have contributed to a greater recognition of the role of occupational factors in the U.S. population burden of COPD. The committee strongly recommends that RDRP support for population-based studies of associations between occupational exposures and COPD continue in order to better define groups of workers at greatest risk and to assist in planning preventive strategies. RDRP efforts to retain spirometry and occupational exposure questions as components of NHANES are critical to better understanding of both the epidemiology of COPD in general and the occupational contribution to the population burden of this disease.

Subgoal: Develop tools and identify at-risk workers in industries and occupations to assess the extent, severity, and burden of work-related COPD.
Recommendation: RDRP work on spirometry, especially the development of new reference equations for normative values, has had a major impact on respira-

tory disease research in general and, more specifically, on preventing COPD. The evaluation committee encourages the RDRP to continue its valuable work on the use of spirometry for longitudinal surveillance of populations known to be at risk for fixed obstructive airway diseases. The committee also recommends that RDRP surveillance activities for work-related COPD and fixed obstructive airways among the general population be established (e.g., there is no SENSOR activity for these conditions). Finally, the committee thought that methods development in the analysis of longitudinal studies of pulmonary function warrants more intense investigations.

Subgoal: Develop, test, and disseminate recommendations for preventing COPD in the workplace.
Recommendation: The RDRP should continue its efforts to support this important subgoal.

Subgoal: Prevent and reduce flavoring-related bronchiolitis obliterans.
Recommendation: The RDRP response to the initial outbreak of diacetyl-induced bronchiolitis obliterans has led to surveillance efforts in multiple locations in an effort to detect and prevent disease. The committee agrees that preventing this disease, both in and of itself and as a model novel disease process, should be a high priority for the RDRP. Because RDRP inhalational toxicologic studies of agents newly recognized to cause airway diseases (e.g., diacetyl) have provided crucial information about mechanisms of disease, the committee strongly recommends that the capacity to conduct such studies be preserved. In addition, because work of the health hazard evaluation (HHE) and technical assistance program was key to identifying both diacetyl and nylon flock as agents that can cause respiratory disease, the RDRP should explore ways to systematically mine data from HHEs that share a common exposure and outcome focus.

STRATEGIC GOAL 2: PREVENT AND REDUCE WORK-RELATED INTERSTITIAL LUNG DISEASES

Much of the RDRP research on interstitial lung diseases is focused on preventing well-known pneumoconioses—coal workers' pneumoconiosis (CWP), silicosis, asbestosis—and chronic beryllium disease (CBD), although the program has contributed greatly to the identification of a disease process caused by exposure to nylon flock. Other newly discovered causes of interstitial lung disease associated with occupational exposures are likely to be identified in the future. The evaluation committee recommendations follow the organizational structure for Chapter 3 of the evidence package.

Subgoal: Coal workers' pneumoconiosis.
Recommendation: NIOSH programs that were established in the 1970s, now under the purview of the RDRP, led to a marked reduction in miners' exposure to respirable coal dust and a decreased annual prevalence of pneumoconiosis through 1999. Despite progress in reducing CWP, surveillance reports have identified hot spots of rapidly progressive CWP in a geographic area that includes eastern Kentucky and southern West Virginia and, more alarmingly, an upward trend in disease prevalence in national data that may be accelerating. On the basis of these new surveillance data, the committee recommends that the RDRP reexamine the organization and efficacy of the CWP surveillance effort, including the interaction between NIOSH and the Mine Safety and Health Administration (MSHA), with additional focus on the adequacy of exposure assessment and compliance determination. The evaluation committee also recommends that the RDRP continue to conduct research to further support MSHA's adoption of the NIOSH recommended exposure limit of 1.0 mg/m^3 as the actual permissible exposure limit for coal mine dust. The committee also recommends research on the possible role of coal rank and silica level in the rapidly progressive cases of CWP as well as other putative risk factors highlighted in NIOSH's hot spot research. Finally, assessment of the strengths and limitations of digital radiography in pneumoconioses in general and its effectiveness in silicons surveillance specifically should be an important continuing research priority.

Subgoal: Silicosis.
Recommendation: The RDRP has made significant contributions to mechanistic understanding of the toxicity of silica and prevention of silicosis. More work in both areas is needed. The evaluation committee recommends that the RDRP continue to support experimental studies of silica-induced cytotoxicity and fibrogenesis and the development of control technologies that include silica substitutes, particle surface coatings, and dust reduction measures. New or overlooked ongoing sources of silica exposure and silicosis should also receive appropriate programmatic attention from NIOSH. As for CWP, assessment of the effectiveness of digital radiography in silicons surveillance should be an important continuing research priority.

Subgoal: Fiber-induced interstitial lung diseases.
Recommendation: The RDRP has played an important role in the epidemiologic evaluation of commercial products that contain asbestos. Despite the well-understood epidemiology of asbestos exposure and significant progress in the control of commercial asbestos exposures, more work needs to be done in this area, especially in terms of understanding fiber-specific cancer risk and the mechanisms of carcinogenesis. Fibers that are asbestiform, such as wincherite and tremolite,

which were found as containments of vermiculite mined in Libby, Montana, or of synthetic origin, such as nylon flock and refractory ceramic fiber, require continued study with attention to fiber characteristics, such as cleavage fragments, and low-level exposures, respectively. The RDRP has also contributed greatly to our understanding of a newly discovered interstitial lung disease among nylon flock workers. While RDRP-assisted control efforts in this industry have been effective, continued surveillance of the industry is needed as well as attention to other, emerging fiber-related respiratory health risks.

Subgoal: Chronic beryllium disease.
Recommendation: The RDRP has made substantial contributions to our understanding of the risks of sensitization to beryllium and progression to CBD. RDRP epidemiologic studies have documented that the Occupational Safety and Health Administration (OSHA) permissible exposure limit does not prevent sensitization or the onset of CBD. Current RDRP efforts are appropriate, but the committee recommends that work be targeted in support of a new recommended OSHA standard that would lead to improved controls to reduce airborne and dermal exposure in all workplaces where beryllium is used. Extending research to other, work-related granulomatous lung diseases is also encouraged.

STRATEGIC GOAL 3: PREVENT AND REDUCE WORK-RELATED INFECTIOUS RESPIRATORY DISEASES

The evidence package presented five intermediate goals, or subgoals, for the infectious disease component of the NIOSH RDRP. These subgoals largely are a continuation of the work being carried out by the RDRP. There is no overarching structure that binds the elements of the infectious disease component into a coherent entity. The work on engineering controls and personal respirator technology runs through the other four elements and could be the unifying theme. Another potential unifying approach would be to structure research activities around the goal of improved understanding of mechanisms of susceptibility to infection resulting from occupational inhalational exposures. An important component that is missing is the need for surveillance activities across known infectious diseases and for emerging threats. In the absence of a unifying theme, the committee recommendations will follow the organizational approach of the evidence package.

Subgoal: Maintain reductions in occupational incidence of tuberculosis (TB) in high-risk work settings.
Recommendation: The goal of the occupational TB program appears to be to prevent and reduce occupationally related TB in the context of the broader response

of the Centers for Disease Control and Prevention to eliminate TB in the United States. This is appropriate, but the lack of specific occupational TB surveillance programs represents a major challenge. Given the continued immigration of documented and undocumented workers from areas with high prevalence of TB, the committee recommends that more specific occupational TB surveillance data be collected, along with exploration of improved methods for TB surveillance.

Subgoal: Protect workers from bioterrorism agents and from occupational acquisition of emerging diseases (including severe acute respiratory syndrome and avian and pandemic flu).
Recommendation: The RDRP played a major role in the national effort to protect workers from the threat of anthrax-contaminated mail in 2001. Improved understanding of how workers became infected from the contaminated mail is needed to increase the "readiness" of government response to a future bioterrorism attack. The RDRP has also been responsive to the specific needs of stakeholders and the public with respect to extramural research on emerging infectious diseases. The committee recommends that research on bioterrorism and emerging infectious diseases be prioritized for extramural funding within the constraints of limited budgetary resources. In addition, NIOSH should coordinate with other federal agencies, such as the Department of Homeland Security, to adopt recently developed technologies for the detection of bioterrorism agents for the protection of workers.

Subgoal: Protect workers from occupational exposures that make them susceptible to respiratory infections.
Recommendation: The RDRP is to be commended for its work on personal respirators and engineering controls for preventing the transmission of infectious agents to workers. The committee recommends that support for these areas remains a high priority.

Subgoal: Prevent outbreaks of occupational histoplasmosis by maintaining worker and employer awareness.
Recommendation: The RDRP histoplasmosis research activity appears to be of historical interest, and the evidence package states that no new RDRP research is planned. In the absence of an operational definition of "awareness" and the documented lack of adequate resources for specific surveillance activities, it is unclear how the RDRP can achieve this subgoal. Given the limited budgetary resources of the agency, the committee recommends that consideration be given to dropping this subgoal.

STRATEGIC GOAL 4: PREVENT AND REDUCE WORK-RELATED RESPIRATORY MALIGNANCIES

Before NORA, the strategic plan for preventing occupational lung diseases did not include lung cancer as a disease target. As a result of the NORA planning process, the RDRP has assumed responsibility for research on lung cancer caused by occupational exposures. The evidence package presented five subgoals relating to determining the occupational etiologies of lung cancer; reducing the incidence of metal-induced lung cancer (hexavalent chromium), silica-induced lung cancer, and lung cancer induced by diesel engine exhaust; and producing lung cancer diagnostic tools.

The impact of the respiratory malignancies program has been strong with regard to the three specific carcinogenic exposures in the workplace listed above—hexavalent chromium, silica, and diesel exhaust—and no specific recommendations are given with respect to these subgoals. In fact, the comprehensive approach to addressing specific carcinogens such as these may serve as a model for developing future initiatives on occupational respiratory malignancies. To aid in future planning, the committee recommends that subgoals with greater specificity for work-related respiratory malignancies be developed to guide the evaluation of research products and productivity.

Subgoal: Determination of occupational etiologies of lung cancer.
Recommendation: The program activities that address the contribution of occupational exposures to the burden of lung cancer in general have led to important work and have partially served as a planning process for addressing respiratory malignancies. However, it is unclear whether separating lung cancer from other important occupationally related cancers is the best way to approach risk, detection, and prevention. The committee recommends that the RDRP ensure that respiratory malignancies are well integrated into an overall program of occupational cancer research and prevention and not arbitrarily segregated from similar research and prevention efforts. In doing so, the RDRP should address the priorities identified in the published report of the NORA1 Task Force on Cancer Research Methods. The committee further recommends that the RDRP develop, in collaboration with other relevant federal agencies (National Cancer Institute [NCI], National Institute of Environmental Health Sciences [NIEHS], Food and Drug Administration, Department of Defense, Department of Labor), a coordinated planning process to identify significant occupational cancer risks in need of a comprehensive approach to detection, surveillance, and prevention.

Subgoal: Development of early diagnostic tools for lung cancer.
Recommendation: The committee questions the relevance and impact of the research into biomarkers for early detection of lung cancer. While this area of investigation is relevant to lung cancer in general, there is little advantage to having this work located at NIOSH instead of at NCI and NIEHS, especially in light of available resources. Focusing the study of such biomarkers on workplace-specific prevention efforts would make these efforts more relevant to NIOSH's mission and would increase the likelihood of the program having a significant impact. The committee recommends that the RDRP consider refocusing the research of biomarkers for early detection to biomarkers of exposure or for early detection that addresses the needs of specific occupational cohorts at high risk of contracting lung cancer.

Recommendation: An emerging issue that could inform research on the risk for work-related respiratory malignancies concerns the inverse relationship between lung cancer and endotoxin exposures. There are a variety of settings with complex exposures to chemical materials where endotoxin exposure occurs, ranging from agricultural settings to metal machining operations. NIOSH should consider whether its expertise in endotoxin positions it to contribute to understanding the role of endotoxin and cancer.

Recommendation: Despite significant contributions to technology for measurement of diesel particulate matter (DPM), NIOSH should continue efforts to develop and validate exposure methods for DPM, especially in the presence of other sources of carbon aerosols, and to provide validation of the methods used to measure DPM in coal mines.

Recommendation: Finally, it is noted that cancers related to asbestos exposure continue to rise. NIOSH should consider developing long-term follow-up studies of exposed workers and interventions, as appropriate, to reduce mortality among these groups. Continued surveillance for asbestos-related risks should receive additional attention.

STRATEGIC GOAL 5: PREVENT RESPIRATORY AND OTHER DISEASES POTENTIALLY RESULTING FROM OCCUPATIONAL EXPOSURES TO NANOMATERIALS

The growing recognition of the usefulness of nanomaterials in various industrial applications has created an urgent need to study the potential health effects of exposures to nanoparticles and methods to control exposures to nanoparticles during manufacturing processes. The RDRP has taken a lead at the national and international levels to address these questions. NIOSH should continue to play a leading role in informing and guiding national and international efforts to address potential occupational hazards and risks associated with the use of manufactured nanoparticles. NIOSH is particularly well suited to have an impact in continuing successful research into methods to monitor exposures to nanomaterials and to develop appropriate engineering controls to prevent such exposures. The following recommendations correspond to the three subgoals listed in the evidence package.

Subgoal: Determine the relative toxicity of nanomaterials.
Recommendation: The RDRP can continue to support some respiratory studies on the toxicology of nanoparticles but because of limited funding should address only those issues that complement studies being supported by other organizations. Thus, there needs to be a continued close interaction with other organizations involved in nanotechnology health-related research.

Subgoal: Conduct exposure assessments and engineering control evaluations in 10 nanomaterial production or use facilities by 2008.
Recommendation: The committee agrees that this is an appropriate and high-priority intermediate goal for the RDRP nanomaterials component.

Subgoal: Produce dose-response data for carbon nanotubes sufficient to conduct a quantitative risk assessment by 2008.
Recommendation: The committee agrees that this is an appropriate intermediate goal but is concerned that data may be insufficient to properly ground a quantitative risk assessment in this short time frame. In particular, data on human health effects are likely to be lacking. The RDRP should recognize that a quantitative risk assessment is not the only approach for dealing with the potential health impact of new materials in a precautionary manner.

CROSS-CUTTING ISSUES

Recommendations Related to Cross-Cutting Issues

During its information-gathering activities and deliberations, the committee identified several issues that cut across all program goals. In Chapter 3, the committee briefly discusses continuing and emerging issues related to surveillance activities, exposure assessment, respirator policy, emergency response, and RDRP resource allocation. The committee provides further discussion and recommendations related to these issues here.

Systems for Surveillance

For all five strategic goals discussed above, a limitation to RDRP research activities is the lack of adequate surveillance methods or data for specific diseases or health outcomes. Chapter 8 of the evidence package (Chapter 8, "Systems for Surveillance") presents information related to RDRP surveillance activities. It is noteworthy that NIOSH has little control over some of the most important surveillance systems on which it depends for some of its most important activities (e.g., NHANES and the National Center for Health Statistics, data for black lung disease). The SENSOR system, although under its control, has limited reach. Even in those areas where it has control and its activities have made important contributions (e.g., CWP), there are gaps in the breadth and depth of the surveillance. In the evidence package, a coherent plan detailing surveillance program needs and necessary resources is not presented, although some limited efforts have been made—for example, with the WoRLD Surveillance Report and the National Surveillance System of Pneumoconiosis Mortality. Articulating such a plan would enable the RDRP to identify places in the NORA2 matrix where the case can be made to compete for resources for such programs. Moreover, such a guiding plan would allow for more rational development and integration of special surveillance systems proposed in the RDRP vision statement into the large data resources of the RDRP and NIOSH.

The success of the HHEs in identifying important occupational health hazards is a notable achievement of the RDRP. In particular, the contribution of HHEs to the recognition of both diacetyl and nylon flock as disease-causing agents underscores the potential value of this data source. However, the evidence package (Chapter 9, "HETA Program Inputs") does not provide a coherent guiding plan as to how the RDRP can make better use of the HHE reports as a surveillance activity.

Recommendation: The committee recommends in the strongest possible terms that NIOSH provide appropriate resources for high-priority surveillance of occupational diseases. The United States is practically alone among highly developed countries with regard to its lack of comprehensive surveillance of occupational diseases. The effectiveness of past NIOSH surveillance activities for coal-dust-related diseases, both CWP and COPD, highlight the importance of improved surveillance for other occupational respiratory disorders.

The RDRP should develop a plan that, at a minimum, includes (1) specific surveillance data needs and gaps in current systems, (2) surveillance unmet currently by any extant system, (3) the specific plans and resource needs required to fill data gaps in items 1 and 2, (4) data integration, and (5) data review to identify new research opportunities. NIOSH should engage in development and evaluation of surveillance methods as a high priority for surveillance of occupational respiratory diseases.

Recommendation: The RDRP should seek to develop a systematic method to review HHEs to mine this data source for identification of new causes of occupational respiratory disorders. Public testimony by RDRP staff indicated that regular review of the HHEs takes place, but no programmatic integration of these activities with the overall surveillance activities and needs of the RDRP was articulated. The value of such articulation in programmatic terms is obvious and should be given high priority by the RDRP.

Exposure Assessment

Exposure assessment within the RDRP incorporates several closely related areas, including (1) development of instrumentation and techniques for sampling and analyzing environmental contaminants in various media, (2) quantitative estimation and modeling of exposure over time from measurement data in support of epidemiologic or risk assessment evaluations, (3) use of existing data sets including administrative data for exposure estimation or surveillance of risk, and (4) estimation of biologically relevant dose by incorporating biomarkers of exposure and understanding the importance of biological processes such as toxicokinetics for peaks and other time-varying exposures. Chapters 2 and 3 describe many excellent examples of relevant and high-impact contributions related to all the program goals.

The evidence package identifies sampling and analytic method development activities as distinct activities not directly linked to the RDRP and states that the RDRP is one of many NIOSH programs that rely on the exposure assessment capabilities. Specific examples of work related to RDRP studies cite the provision of sampling and analytic "support" for the field studies undertaken by RDRP scientists. Thus, the evidence package refers frequently to exposure assessment activities as services provided to the RDRP research activities rather than as an integral component of the research questions. The RDRP does not present these exposure assessment activities as part of an explicit or comprehensive focus on exposure assessment methods. The implication of this presentation is that, although exposure assessment is an important part of RDRP studies, exposure assessment scientists are not necessarily an integral part of the research effort. Further, while a number of RDRP scientists have provided a focus on exposure assessment research in the past, no specific mention of current or future staffing and programmatic needs for exposure assessment activities is made explicitly.

Recommendation: The committee recommends strongly that the RDRP develop a programmatic goal that provides an explicit or comprehensive focus on exposure assessment methods.

Emergency Response

The RDRP has made important contributions to the emergency response to recent disasters, including the World Trade Center and anthrax terror attacks and hurricanes Katrina and Rita. However, there is no articulation of how this emergency response work is integrated into the framework of the RDRP research activities conducted under the five strategic goals, and it is unclear whether the RDRP plans to incorporate research activities related to emergency responses into future efforts. Much could be learned about exposure-response relationships and ultimately protecting emergency responders by conducting longitudinal cohort studies relating to catastrophic events. Information from emergency responses to toxicant exposures should be applicable to models of irritant-induced asthma, fixed airway obstruction, interstitial lung disease, and possibly even malignancies.

Recommendation: The RDRP is encouraged to include research strategies in emergency response efforts. For example, the RDRP should assess how NIOSH-supported research and medical surveillance of World Trade Center disaster emergency responders and recovery workers may or may not be relevant to WRA, work-related fixed obstructive airway disease, interstitial lung disease, and possibly malignancies.

RDRP Resource Allocation

The RDRP was created during the recent NORA2 reorganization of NIOSH. NORA2 emphasizes industry-sector-based research as opposed to the previous organization, which emphasized disease-based research. As such, the emphasis on respiratory diseases is clearly no longer as targeted. Questions remain about how research priorities for respiratory diseases that cut across sectors will be treated and whether the quality and impact of RDRP research will remain. In addition, because NIOSH has limited funds available for extramural awards of any type, the RDRP needs a system(s) to ensure that unnecessary duplication and inappropriate expenditure on low-priority research projects are avoided.

Recommendation: The committee recommends that careful regular review of programmatic goals and achievement of those goals be conducted to ensure that the respiratory disease components are being maintained under the matrix management structure of NORA2. All research proposals that are under consideration for funding, whether intramural or extramural, should be prioritized according to the RDRP strategic plan, which needs to be updated periodically.

BROADER PROGRAMMATIC GOALS

Broader programmatic goals (visionary issues) have also been developed to assist the RDRP in identifying and effectively addressing issues that may arise in the future.

- Research coordination: The RDRP should work closely with NIEHS; the National Heart, Lung, and Blood Institute; the Environmental Protection Agency; and other relevant federal agencies to develop a truly coordinated approach to research on occupational and environmental respiratory disorders. Current institutional silos obstruct the efficient use of resources and development of knowledge.
- Surveillance: The RDRP should develop surveillance systems for occupational and environmental respiratory disorders that generate adequate prevalence and incidence data for the appropriate targeting of resources for preventing and reducing the disease burden. Attention to developing systems that could capture evidence of unrecognized risks and new outbreaks is also needed.
- Outreach to the pulmonary research community: The RDRP has played an important role in the past to focus and catalyze research efforts on important occupational exposures and work-related lung diseases by hosting or cosponsoring conferences and workshops. The evaluation committee strongly recommends continued support of such activities to stimulate extramural investigators to target priority issues identified by the RDRP strategic-planning process.

References

Agrawal, A., J. Cronin, J. Tonazzi, T.M. McCleskey, D.S. Ehler, E.M. Minogue, G. Whitney, C. Brink, A.K. Burrell, B. Warner, M.J. Goldcamp, P.C. Schlecht, P. Sonthalia, and K. Ashley. 2006. Validation of a standardized portable fluorescence method for determining trace beryllium in workplace air and wipe samples. J. Environ. Monit. 8(6):619-624.

AIA/FGI (American Institute of Architects/Facility Guidelines Institute). 2006. Guidelines for Design and Construction of Health Care Facilities: 2006 Edition: Facilities Guidelines Institute and the AIA Academy of Architecture for Health, with assistance from the U.S. Department of Health and Human Services.

Aizenberg, V., E. England, S. Grinshpun, K. Willeke, and G. Carlton. 2000. Metal exposure among abrasive blasting workers at four U.S. air force facilities. Appl. Occup. Environ. Hyg. 15(10):766-772.

Akbar-Khanzadeh, F., S. Milz, A. Ames, P.P. Susi, M. Bisesi, S.A. Khuder, and M. Akbar-Khanzadeh. 2007. Crystalline silica dust and respirable particulate matter during indoor concrete grinding—wet grinding and ventilated grinding compared with uncontrolled conventional grinding. J. Occup. Environ. Hyg. 4(10):770-779.

American Lung Association. 2006. Chronic Obstructive Pulmonary Disease (COPD) Fact Sheet: (Chronic Bronchitis and Emphysema). August 2006 [online]. Available: http://www.lungusa.org/site/pp.asp?c=dvLUK9O0E&b=35020 [accessed Feb.15, 2008].

Antao, V., and C. Piacitelli. 2004. Health Hazard Evaluation Report: Claremont Flock Corporation, Leominster, Massachusetts. HETA-2004-0186-3011. U.S. Department of Health and Human Services, Centers for Disease Control, National Institute for Occupation Safety and Health, Cincinnati, OH [online]. Available: http://www.cdc.gov/niosh/hhe/reports/pdfs/2004-0186-3011.pdf [accessed Aug. 15, 2007].

Antao, V.C., E.L. Petsonk, L.Z. Sokolow, A.L. Wolfe, G.A. Pinheiro, J.M. Hale, and M.D. Attfield. 2005. Rapidly progressive coal workers' pneumoconiosis in the United States: Geographic clustering and other factors. Occup. Environ. Med. 62(10):670-674.

Antonini, J.M., J.R. Roberts, R.W. Clarke, H.M. Yang, M.W. Barger, J.Y. Ma, and D.N. Weissman. 2001a. Effect of age on respiratory defense mechanisms: Pulmonary bacterial clearance in Fischer 344 rats after intratracheal instillation of *Listeria monocytogenes.* Chest 120(1):240-249.

Antonini, J.M., H.M. Yang, J.Y. Ma, J.R. Roberts, M.W. Barger, L. Butterworth, T.G. Charron, and V. Castranova. 2001b. Subchronic silica exposure enhances respiratory defense mechanisms and the pulmonary clearance of *Listeria monocytogenes* in rats. Inhal. Toxicol. 12(11):1017-1036.

Antonini, J.M., M.D. Taylor, L. Mileecchia, A.R. Bebout, and J.R. Roberts. 2004. Suppression in lung defense responses after bacterial infection in rats pretreated with different welding fumes. Toxicol. Appl. Pharmacol. 200(3):206-218.

Antonini, J.M., A.A. Afshari, S. Stone, B. Chen, D. Schwegler-Bery, G.W. Fletcher, T.W. Goldsmith, K.H. Vandestouwe, W. McKinney, V. Castranova, and D.G. Frazer. 2006. Design, construction, and characterization of a novel robotic welding fume generator and inhalation exposure system for laboratory animals. J. Occup. Environ. Hyg. 3(4):194-203.

Ashley, K., A.M. Howe, M. Demange, and O. Nygren. 2003. Sampling and analysis considerations for the determination of hexavalent chromium in workplace air. J. Environ. Monit. 5(5):707-716.

Astrakianakis, G., N.S. Seixas, R. Ray, J.E. Camp, D.L. Gao, Z. Feng, W. Li., K.J. Wernli, E.D. Fitzgibbons, D.B. Thomas, and H. Checkoway. 2007. Lung cancer risk among female textile workers exposed to endotoxin. J. Natl. Cancer Inst. 99(5):357-364.

Attfield, M.D., and J. Costello. 2004. Quantitative exposure-response for silica dust and lung cancer in Vermont granite workers. Am. J. Ind. Med. 45(2):129-138.

Attfield, M.D., and T.K. Hodous. 1992. Pulmonary function of U.S. coal miners related to dust exposure estimates. Am. Rev. Respir. Dis. 145(3):605-609.

Attfield, M.D., and K. Morring. 1992a. The derivation of estimated dust exposures for U.S. coal miners working before 1970. Am.Ind. Hyg. Assoc. J. 53(4):248-255.

Attfield, M.D., and K. Morring. 1992b. An investigation into the relationship between coal workers' pneumoconiosis and dust exposure in U.S. coal miners. Am. Ind. Hyg. Assoc. J. 53(8):486-492.

Attfield, M.D., and N.S. Seixas. 1995. Prevalence of pneumoconiosis and its relationship to dust exposure in a cohort of U.S. bituminous coal miners and ex-miners. Am. J. Ind. Med. 27(1):137-151.

Balmes, J., M. Becklake, P. Blanc, P. Henneberger, K. Kreiss, C. Mapp, D. Milton, D. Schwartz, K. Toren, and G. Viegi. 2003. American Thoracic Society statement: Occupational contributions to the burden of airway disease. Am. J. Respir. Crit. Care. Med. 167(5):787-797.

Baron, P., R. Castellan, V. Castronova, D. Dankovic, C. Doak, D. Lewis, L. Stettler, K. Weber, and R. Zumwalde. 1999. NIOSH Interdivisional Fiber Subcommittee Final report, March 17, 1999.

Beeckman, L.F., M.L. Wang, E.L. Petsonk, and G.R. Wagner. 2001. Rapid declines in FEV1 and subsequent respiratory symptoms, illnesses, and mortality in coal miners in the U.S. Am. J. Respir. Crit. Care Med. 163(3 Pt 1):633-639.

Belle, G.K., R.V. Ramani, and J.F. Colinet. 2000. Evaluation of two-phase spray system for airborne dust control in a longwall gallery. Pp. 113-119 in Proceedings of 12th International Conference on Coal Research, Sandton, Republic of South Africa, Sep. 12-15, 2000. Washington, DC: International Conference on Coal Research [online]. Available: http://www.cdc.gov/niosh/mining/pubs/pdfs/eotsp.pdf [accessed Aug. 15, 2007].

Birch, M.E., and J.D. Noll. 2004. Submicrometer elemental carbon as a selective measure of diesel particulate matter in coal mines. J. Environ. Monit. 6(10):799-806.

Blake, T., V. Castranova, D. Schwegler-Berry, P. Baron, G.J. Deye, C. Li, and W. Jones. 1998. Effect of fiber length on glass microfiber cytotoxicity. J. Toxicol. Environ. Health 54(4):243-259.

Blanc, P.D., and K. Torén. 2007. Occupation in chronic obstructive pulmonary disease and chronic bronchitis: An update. Int. J. Tuberc. Lung Dis. 11(3):251-257.

Boag, A.H., T.V. Colby, A.E. Fraire, C. Kuhn III, V.L. Roggli, W.D. Travis, and V. Vallyathan. 1999. The pathology of interstitial lung disease in nylon flock workers. Am. J. Surg. Pathol. 23(12):1539-1545.

Boeniger, M.F., Z.L. Lummus, R.E. Biagini, D.I. Bernstein, M.C. Swanson, C. Reed, and M. Massoudi. 2001. Exposure to protein aeroallergens in egg processing facilities. Appl. Occup. Environ. Hyg. 16(6):660-670.

Boffetta, P. 2007. Endotoxins in lung cancer prevention. J. Natl. Cancer Inst. 99(5):339.

Boiano, J.M., M.E. Wallace, W.K. Sieber, J.H. Groff, J. Wang, and K. Ashley. 2000. Comparison of three sampling and analytical methods for the determination of airborne hexavalent chromium. J. Environ. Monit. 2(4):329-333.

Boylstein, R., C. Piacitelli, A. Grote, R. Kanwal, G. Kullman, and K. Kreiss. 2006. Diacetyl emissions and airborne dust from butter flavorings used in microwave popcorn production. Appl. Occup. Environ. Hyg. 3(1):530-535.

Breton, C.V., Z. Zhang, P.R. Hunt, E. Pechter, and L. Davis. 2006. Characteristics of work related asthma: Results from a population based survey. Occup. Environ. Med. 63(6):411-415.

Bugarski, A.D. 2004. Characterization of diesel aerosols in an underground metal mine. Proceedings of the 8th ETH Conference on Combustion Generated Nanoparticles, August 16-18, 2004, Zurich, Switzerland, A. Mayer, ed. Zurich: Swiss Federal Institute of Technology.

Bugarski, A., and M. Gautam. 2001. Size distribution and deposition in human respiratory tract: Particle mass and number. Proceedings of the 4th International ETH-Conference on Nanoparticle Measurement, August 7-9, 2000, Bern, Switzerland, A. Mayer, ed. Zurich: Swiss Federal Institute of Technology [online]. Available: http://www.cdc.gov/niosh/mining/pubs/pdfs/sdadi.pdf [accessed Aug. 15, 2007].

Bugarski, A.D., and G.H. Schnakenberg, Jr. 2001. Field Evaluation of Diesel Particulate Filters at Brunswick Mine. Presentation at the Mining Diesel Emissions Conference, November 7-8, 2001, Markham, Ontario.

Bugarski, A.D., and G.H. Schnakenberg, Jr. 2003. Testing of Diesel Emissions Control Technologies in Isolated Zone. Presentation at the Mining Diesel Emissions Conference (MDEC), November 4-6, 2003, Markham, Ontario.

Bugarski, A., G.H. Schnakenberg, Jr., J.D. Noll, S. Mischler, M.W. Crum, and R. Anderson. 2004a. Evaluation of Diesel Particulate Filter Systems and Biodiesel Blends in an Underground Mine. SME Annual Meeting, February 23-25, 2004, Denver, CO. SME preprint 04-24. Littleton, CO: Society for Mining, Metallurgy, and Exploration, Inc.

Bugarski, A.D., G.H. Schnakenberg, Jr., J.D. Noll, S. Mischler, L. Patts, J. Hummer, S. Vanderslice, M. Crum, and R. Anderson. 2004b. The Effectiveness of Selected Technologies in Controlling Diesel Emissions in an Underground Mine—Isolated Zone Study at Stillwater Mining Company's Nye Mine. Final Report to Metal/Nonmetal Diesel Partnership January 5, 2004 [online]. Available: http://www.msha.gov/01-995/dpmdocs/stillwater.pdf [accessed June 27, 2007].

Bugarski, A.D., S.E. Mischler, and G.H. Schnakenberg, Jr. 2005. Effects of alternative fuels on concentrations of nanometer and ultrafine particles in underground mine. Proceedings of the 9th ETH Conference on Combustion Generated Nanoparticles, August 15-17, 2005, Zurich, Switzerland, A. Mayer, ed. Zurich: Swiss Federal Institute of Technology.

Bugarski, A., G.H. Schnakenberg, Jr, J.D. Noll, S.E. Mischler, M.W. Crum, and R. Anderson. 2006. Evaluation of diesel particulate filter systems and biodiesel blends in an underground mine. Trans. Soc. Min. Metal Explor. 318:27-35.

Burkhart, J., C. Piacitelli, D. Schwegler-Berry, and W. Jones. 1999. Environmental study of nylon flocking process. J. Toxicol. Environ. Health A 57(1):1-23.

CalEPA (California Environmental Protection Agency). 1998. Report to the Air Resources Board on the Proposed Identification of Diesel Exhaust as a Toxic Air Contaminant, Part A: Exposure Assessment As Approved by the Scientific Review Panel on April 22, 1998. California Environmental Protection Agency [online]. Available: http://www.arb.ca.gov/toxics/dieseltac/part_a.pdf [accessed Aug. 21, 2007].

Campoli, A.A., F.E. McCall, G.L. Finfinger, and M.D. Zuber. 1996. Longwall dust control potentially enhanced by surface borehole water infusion. Min. Eng. 48(7):56-60.

Cantrell, B.K., S.W. Stein, H. Patashnick, and D. Hassel. 1996. Status of a tapered element, oscillation microbalance-based continuous respirable coal mine dust monitor. Appl. Occup. Environ. Hyg. 11(7):624-629.

Castranova, V. 2004. Signaling pathways controlling the production of inflammatory mediators in response to crystalline silica exposure: Role of reactive oxygen/nitrogen species. Free Radic. Biol. Med. 37(7):916-925.

Castranova, V., V. Vallyathan, and W.E. Wallace, eds. 1996a. Silica and Silica-Induced Lung Diseases. Boca Raton: CRC Press.

Castranova, V., W.H. Pailes, N.S. Dalal, P.R. Miles, L. Bowman, V. Vallyathan, D. Pack, K.C. Weber, A. Hubbs, D. Schwegeler-Berry, J. Xiang, R. Dey, J. Blackford, J.Y.C. Ma, M. Barger, D.A. Shoemaker, J.R. Pretty, D.M. Ramsey, J.L. McLaurin, A. Klan, P.A. Baron, C.P. Childress, L.E. Stettler, and A.W. Teass. 1996b. Enhanced pulmonary response to the inhalation of freshly fractured silica as compared to aged dust exposure. Appl. Occup. Environ. Hyg. 11(7):937-941.

CDC (Centers for Disease Control and Prevention). 1988. Carcinogenic Effects of Exposure to Diesel Exhaust: Recommendations. Current Intelligence Bulletin 50. August 1988 [online]. Available: www.cdc.gov/niosh/88116_50.html#Recommendations [accessed June 27, 2007].

CDC (Centers for Disease Control and Prevention). 2000. Healthy People 2010: Cancer. U.S. Department of Health and Human Services, Centers for Disease Control and Prevention, National Institute of Health. November 2000 [online]. Available: www.healthypeople.gov/Document/pdf/Volume1/03Cancer.pdf [accessed June 27, 2007].

CDC (Centers for Disease Control and Prevention). 2001. CDC Interim: Recommendations for Protecting Workers from Exposure to *Bacillus anthracis* in Work Sites Where Mail Is Handled or Processed. Health Alert Network: October 31, 2001 (Updated from CDC Health Advisory 45 issued 10/24/01). U.S. Department of Health and Human Services, Centers for Disease Control and Prevention, Atlanta, GA [online]. Available: http://www2a.cdc.gov/HAN/ArchiveSys/ViewMsgV.asp?AlertNum=00051 [accessed Aug. 16, 2007].

CDC (Centers for Disease Control and Prevention). 2002. CDC's Response to Ending Neglect: The Elimination of Tuberculosis in the United States. U.S. Department of Health and Human Services, Centers for Disease Control and Prevention, Atlanta, GA [online]. Available: http://www.cdc.gov/tb/pubs/iom/iomresponse/iomresponse.pdf [accessed Aug. 17, 2007].

CDC (Centers for Disease Control and Prevention). 2003. Influenza (Flu): Respiratory Hygiene/Cough Etiquette in Healthcare Settings. U.S. Department of Health and Human Services, Centers for Disease Control and Prevention [online]. Available: http://www.cdc.gov/flu/professionals/infectioncontrol/resphygiene.htm [accessed Aug. 22, 2007].

CDC (Centers for Disease Control and Prevention). 2004a. Public Health Guidance for Community-Level Preparedness and Response to Severe Acute Respiratory Syndrome (SARS) Version 2—Supplement I: Infection Control in Healthcare, Home, and Community Settings. U.S. Department of Health and Human Services, Centers for Disease Control and Prevention [online]. Available: http://www.cdc.gov/NCIDOD/SARS/guidance/i/pdf/i.pdf [accessed Aug. 17, 2007].

CDC (Centers for Disease Control and Prevention). 2004b. Guidance about SARS for Airline Flight Crews, Cargo and Cleaning Personnel, and Personnel Interacting with Arriving Passengers. Severe Acute Respiratory Syndrome. U.S. Department of Health and Human Services, Centers for Disease Control and Prevention. April 23, 2004 [online]. Available: http://www.cdc.gov/ncidod/sars/pdf/airpersonnel.pdf [accessed Aug. 17, 2007].

CDC (Centers for Disease Control and Prevention). 2004c. Avian Influenza (Flu): Interim Recommendations for Infection Control in Health-Care Facilities Caring for Patients with Known or Suspected Avian Influenza. U.S. Department of Health and Human Services, Centers for Disease Control and Prevention [online]. Available: http://www.cdc.gov/flu/avian/professional/infect-control.htm [accessed Aug. 22, 2007].

CDC (Centers for Disease Control and Prevention). 2007a. Antrax. Emergency Preparedness and Response. U.S. Department of Health and Human Services, Centers for Disease Control and Prevention [online]. Available: http://www.bt.cdc.gov/agent/anthrax/ [accessed Aug. 22, 2007].

CDC (Centers for Disease Control and Prevention). 2007b. Avian Influenza (Bird Flu). U.S. Department of Health and Human Services, Centers for Disease Control and Prevention [online]. Available: http://www.cdc.gov/flu/avian/index.htm [accessed Aug. 22, 2007]

Christian, M.D., M. Loutfy, L.C. McDonald, K.F. Martinez, M. Ofner, T. Wong, T. Wallington, W.L. Gold, B. Mederski, K. Green, and D.E. Low. 2004. Possible SARS coronavirus transmission during cardiopulmonary resuscitation. Emerg. Infect. Dis. 10(2):287-293.

Coffey, C.C., R.B. Lawrence, D.L. Campbell, Z. Zhuang, C.A. Calvert, and P.A. Jensen. 2004. Fitting characteristics of eighteen N95 filtering-facepiece respirators. J. Occup. Environ. Hyg. 1(4):262–271.

Coffey, C.C., R.B. Lawrence, Z. Zhuang, M.G. Duling, and D.L. Campbell. 2006. Errors associated with three methods of assessing respirator fit. J. Occup. Environ. Hyg. 3(1):44-52.

Cohen, M.D., M. Sisco, K. Baker, L.C. Chen, and R.B. Schlesinger. 2002. Rapid communication: Effect of inhaled chromium on pulmonary A1AT. Inhal. Toxicol. 14(7):765-771.

Costello, J., and W.G. Graham. 1988. Vermont granite workers' mortality study. Am. J. Ind. Med. 13(4):483-497.

Craighead, J.E., J. Kleinerman, J.F. Abraham, A.R. Gibbs, N.V. Vallyathan, and E.B. Juliano. 1988. Diseases associated with exposure to silica and non-fibrous silicate minerals. Arch. Pathol. Lab. Med. 112(7):673-720.

Croteau, G.A., M.E. Flanagan, J.E. Camp, and N.S. Seixas. 2004. The efficacy of local exhaust ventilation for controlling dust exposures during concrete surface grinding. Ann. Occup. Hyg. 48(6):509-518.

Cummings, K.J., D.C. Deubner, G.A. Day, P.K. Henneberger, M.M. Kitt, M.S. Kent, K. Kreiss, and C.R. Schuler. 2007. Enhanced preventive programme at a beryllium oxide ceramics facility reduces beryllium sensitization among new workers. Occup. Environ. Med. 64(2):134-140.

Daroowalla, F., M.L. Wang, C. Piacitelli, J. Burkhart, and W. Jones. 1998a. Health Hazard Evaluation Report: Claremont Flock Corporation, Claremont, NH. HETA-98-0212-2788. U.S. Department of Health and Human Services, Public Health Service, Centers for Disease Control and Prevention, National Institute for Occupational Safety and Health, Cincinnati, OH [online]. Available: http://www.cdc.gov/niosh/hhe/reports/pdfs/1998-0212-2788.pdf [accessed Aug. 15, 2007].

Daroowalla, F., M.L. Wang, C. Piacitelli, J. Burkhart, W. Jones. 1998b. Health Hazard Evaluation Report: Spectro Coating Corporation, Leominster, Massachusetts. HETA-98-0238-2789. U.S. Department of Health and Human Services, Public Health Service, Centers for Disease Control and Prevention, National Institute for Occupational Safety and Health, Cincinnati, OH [online]. Available: http://www.cdc.gov/niosh/hhe/reports/pdfs/1998-0238-2789.pdf [accessed Aug. 15, 2007].

Daroowalla, F., M.L. Wang, C. Piacitelli, M.D. Attfield, and K. Kreiss. 2005. Flock workers' exposures and respiratory symptoms in five plants. Am. J. Ind. Med. 47(2):144-152.

Day, G.A., A. Dufresne, A.B. Stefaniak, C.R. Schuler, M.L. Stanton, W.E. Miller, M.S. Kent, D.C. Deubner, K. Kreiss, and M.D. Hoover. 2007. Exposure pathway assessment at a copper-beryllium alloy facility. Ann. Occup. Hyg. 51(1):67-80

Deye, G.J., P. Gao, P.A. Baron, and J. Fernback. 1999. Performance of a fiber length classifier. Aerosol Sci. Technol. 30(5):420-437.

DHHS (U.S. Department of Health and Human Services). 2006. The Health Consequences of Involuntary Exposure to Tobacco Smoke: A Report of the Surgeon General. U.S. Department of Health and Human Services. June 27, 2006 [online]. Available: http://www.surgeongeneral.gov/library/secondhandsmoke/ [accessed Nov. 26, 2007].

Dillard, S.F., B. Hefflin, R.G. Kaczmarek, E.L. Petsonk, and T.P. Gross. 2002. Health effects associated with medical glove use. AORN J. 76(1):88-96.

Ding, M., X. Shi, V. Castranova, and V. Vallyathan. 2000. Predisposing factors in occupational lung cancer: inorganic minerals and chromium. J. Environ. Pathol. Toxicol. Oncol. 19(1-2):129-138.

Donovan Reh, B. 2004. A Summary of Health Hazard Evaluations: Issues Related to Occupational Exposure to Isocyanates, 1989 to 2002. DHHS (NIOSH) Publication No. 2004-116. U.S. Department of Health and Human Services, Centers for Disease Control and Prevention, National Institute for Occupational Safety and Health. January 2004 [online]. Available: http://www.cdc.gov/niosh/docs/2004-116/pdfs/2004-116.pdf [accessed Nov. 14, 2007].

Driscoll, T., D.I. Nelson, K. Steenland, J. Leigh, M. Concha-Barrientos, M. Fingerhut, and A. Prüss-Üstün. 2005a. The global burden of non-malignant respiratory disease due to occupational airborne exposures. Am. J. Ind. Med. 48(6):432-445.

Driscoll, T., D.I. Nelson, K. Steenland, J. Leigh, M. Concha-Barrientos, M. Fingerhut, and A. Prüss-Ustün. 2005b. The global burden of disease due to occupational carcinogens. Am. J. Ind. Med. 48(6):419-431.

Dunn, K.H., S.A. Shulman, A.B. Cecala, and D.E. Ventuirin. 2004. Evaluation of a local exhaust ventilation system for controlling refractory ceramic fibers during disc sanding. J. Occup. Environ. Hyg. 1(10):D107-D111.

Echt, A., and W.K. Sieber. 2002. Control of silica exposures from hand tools in construction: Grinding concrete. Appl. Occup. Environ. Hyg. 17(7):457-461.

Echt, A., W. Sieber, A. Jones, and E. Jones. 2002. Control of silica exposure in construction: Scabbling concrete. Appl. Occup. Environ. Hyg. 17(12):809-813.

Echt, A., W. Sieber, E. Jones, D. Schill, D. Lefkowitz, J. Sugar, and K. Hoffner. 2003. Control of respirable dust and crystalline silica from breaking concrete with a jackhammer. Appl. Occup. Environ. Hyg. 18(7):491-495

Enterline, P.E., J.L. Sykora, G. Keleti, and J.H. Lange. 1985. Endotoxin, cotton dust and cancer. Lancet 2(8461):934-935.

EPA (U.S. Environmental Protection Agency). 2002. Health Assessment Document for Diesel Engine Exhaust. EPA/600/8-90/057F. National Center for Environmental Assessment, Office of Research and Development, U.S. Environmental Protection Agency, Washington, DC [online]. Available: http://cfpub.epa.gov/ncea/cfm/recordisplay.cfm?deid=29060 [accessed July 26, 2007].

Eschenbacher, W.L., K. Kreiss, D. Lougheed, G.S. Pransky, B. Day, and R.M. Castellan. 1999. Nylon flock-associated interstitial lung disease. Am. J. Respir. Crit. Care Med. 159(6):2003-2008.

Fedan, J.S., J.A. Dowdy, K.B. Fedan, and A.F. Hubbs. 2006. Popcorn worker's lung: In vitro exposure to diacetyl, an ingredient in microwave popcorn butter flavoring, increases reactivity to methacholine. Toxicol. Appl. Pharmacol. 215(1):17-22.

Feldman, D.M., S.L. Baron, B.P. Bernard, B.D. Lushniak, G. Banauch, N. Arcentales, K.J. Kelly, and D.J. Prezant. 2004. Symptoms, respirator use, and pulmonary function changes among New York City firefighters responding to the World Trade Center disaster. Chest 125(4):1256-1264.

Fennelly, K.P., J.W. Martyny, K.E. Fulton, I.M. Orme, D.M. Cave, and L.B. Heifets. 2004. Cough-generated aerosols of *Mycobacterium tuberculosis*: A new method to study infectiousness. Am. J. Respir. Crit. Care Med. 169(5):604-609.

Flanagan, M.E., N. Seixas, M. Majar, J. Camp, and M. Morgan. 2003. Silica dust exposures during selected construction activities. AIHA J. 64(3):319-328.

Flattery, J., L. Davis, K.D. Rosenman, R. Harrison, S. Lyon-Callo, and M. Filios. 2006. The proportion of self-reported asthma associated with work in three states: California, Massachusetts, and Michigan, 2001. J. Asthma 43(3):213-218.

Flynn, M.R., and P. Susi. 2003. Engineering controls for selected silica and dust exposures in the construction industry—a review. Appl. Occup. Environ. Hyg. 18(4):268-277.

Gautrin, D., I.L. Bernstein, S. Brooks, and P. Henneberger. 2006. Reactive airways dysfunction syndrome and irritant-induced asthma. Pp. 581-629 in Asthma in the Workplace, 3rd Ed., I.L. Bernstein, M. Chan-Yeung, J.L. Malo, and D.I. Bernstein, eds. New York: Taylor & Francis.

Goe, S.K., P.K. Henneberger, M.J. Reilly, K.D. Rosenman, D.P. Schill, D. Valiante, J. Flattery, R. Harrison, F. Reinisch, C. Tumpowsky, and M.S. Filios. 2004. A descriptive study of work aggravated asthma. Occup. Environ. Med. 61(6):512-517.

Green, F.H., R. Althouse, and K.C. Weber. 1989. Prevalence of silicosis at death in underground coal miners. Am. J. Ind. Med. 16(6):605-616.

Gulumian, M., P.J. Borm, V. Vallyathan, V. Castranova, K. Donaldson, G. Nelson, and J. Murray. 2006. Mechanistically identified suitable biomarkers of exposure, effect, and susceptibility for silicosis and coal-worker's pneumoconiosis: A comprehensive review. J. Toxicol. Environ. Health B Crit. Rev. 9(5):357-395.

Hall, R.M. and B.P. Bernard. 2002. Letter to Claudio Benedi, Publications Service Division, U.S. Office of Personnel Management, Washington, DC, from Ronald M. Hall, and Bruce P. Bernard, Hazard Evaluations and Technical Assistance Branch, National Institute for Occupational Safety and Health, Cincinnati, OH. HETA 2002-0114. August 30, 2002.

Hall, R.M. and J.E. Hess. 2002. Letter to Sam Pulcrano, Safety Performance Management, U.S. Postal Service, Washington, DC, from Ronald M. Hall, and Jeffery E. Hess, Hazard Evaluations and Technical Assistance Branch, National Institute for Occupational Safety and Health, Cincinnati, OH. HETA 2002-0079. March 14, 2002.

Hall, R.M., B. Bernard, J. Harney, R. McCleery, M. Gillen, J.E. Hess, M. Kiefer, D. Mattorano, L. Delaney, and K. Mead. 2002. U.S. Senate and House of Representatives, Washington, DC, July 2002. NIOSH Health Hazard Evaluation Report HETA 2002-0136-2880.

Hankinson, J.L., J.R. Odencrantz, and K.B. Fedan. 1999. Spirometric reference values from a sample of the general U.S. population. Am. J. Respir. Crit. Care Med. 159(1):179-187.

Henderson, V., and P.E. Enterline. 1973. An unusual mortality experience in cotton textile workers. J. Occup. Med. 15(9):717-719.

Henneberger, P.K., K. Kreiss, K.D. Rosenman, M.J. Reilly, Y.F. Chang, and C.A. Geidenberger. 1999. An evaluation of the incidence of work-related asthma in the United States. Int. J. Occup. Med. Environ. Health 5(1):1-8.

Henneberger, P.K., D. Cumro, D.D. Deubner, M.S. Kent, M. McCawley, and K. Kreiss. 2001. Beryllium sensitization and disease among long-term and short-term workers in a beryllium ceramics plant. Int. Arch. Occup. Environ. Health 74(3):167-176.

Henneberger, P.K., R.D. Deprez, N. Asdigian, L.C. Oliver, S. Derk, and S.K. Goe. 2003a. Workplace exacerbation of asthma symptoms: Findings from a population-based study in Maine. Arch. Environ. Health 58(12):781-788.

Henneberger, P.K., S.J. Derk, L. Davis, C. Tumpowsky, M.J. Reilly, K.D. Rosenman, D.P. Schill, D. Valiante, J. Flattery, R. Harrison, F. Reinisch, M.S. Filios, and B. Tift. 2003b. Work-related reactive airways dysfunction syndrome cases from surveillance in selected US states. J. Occup. Environ. Med. 45(4):360-368.

Henneberger, P.K., S.K. Goe, W.E. Miller, B. Doney, and D.W. Groce. 2004. Industries in the U.S. with airborne beryllium exposure and estimates of the number of current workers potentially exposed. J. Occup. Environ. Hyg. 1(10):648-659.

Henneberger, P.K., S.J. Derk, S.R. Sama, R.J. Boylstein, C.D. Hoffman, P.A. Preusse, R.A. Rosiello, and D.K. Milton. 2006. The frequency of workplace exacerbation among health maintenance organization members with asthma. Occup. Environ. Med. 63(8):551-557.

Hnizdo, E., and D.C. Sylvain. 2003. Health Hazard Evaluation Report: ChemDesign Corporation, Fitchburg, MA. HETA-2000-0096-2876. U.S. Department of Health and Human Services, Centers for Disease Control, National Institute for Occupation Safety and Health, Cincinnati, OH [online]. Available: http://www.cdc.gov/niosh/hhe/reports/pdfs/2000-0096-2876.pdf [accessed Nov, 18, 2007].

Hnizdo, E., and V. Vallyathan. 2003. Chronic obstructive pulmonary disease due to occupational exposure to silica dust: A review of epidemiological and pathological evidence. Occup. Environ. Med. 60(4):237-243.

Hnizdo, E., P.A. Sullivan, K.M. Bang, and G. Wagner. 2002. Association between chronic obstructive pulmonary disease and employment by industry and occupation in the US population: A study of data from the Third National Health and Nutrition Examination Survey. Am. J. Epidemiol. 156(8):738-746.

Hnizdo, E., D. Sylvain, D.M. Lewis, E. Pechter, and K. Kreiss. 2004a. New onset asthma associated with exposure to 3-amino-5-mercapto-1,2,4-triazole. J. Occup. Environ. Med. 46(12):1246-1252.

Hnizdo, E., P.A. Sullivan, K.M. Bang, G. Wagner. 2004b. Airflow obstruction attributable to work in industry and occupation among U.S. race/ethnic groups: A study of NHANES III data. Am. J. Ind. Med. 46(2):126-135.

Hnizdo, E., L. Yu, L. Freyder, M.D. Attfield, J. Lefante, and H.W. Glindmeyer. 2005. The precision of longitudinal lung function measurements: Monitoring and interpretation. Occup. Environ. Med. 62(10):695-701.

Hnizdo, E., K. Sircar, H.W. Glindmeyer, and E.L. Petsonk. 2006a. Longitudinal limits of normal decline in lung function in an individual. J. Occup. Environ. Med. 48(6):625-634.

Hnizdo, E., H.W. Glindmeyer, E.L. Petsonk, P. Enright, and A.S. Buist. 2006b. Case definitions for chronic obstructive pulmonary disease. COPD 3(2):95-100.

Hodgins, P., P.K. Henneberger, M.L. Wang, and E.L. Petsonk. 1998. Bronchial responsiveness and five-year FEV1 decline: A study in miners and nonminers. Am. J. Respir. Crit. Care Med. 157(5 Pt. 1):1390-1396.

Hodgson, J.T., and R.D. Jones. 1990. Mortality of workers in the British cotton industry in 1968-1984. Scand. J. Work Environ. Health 16(2):113-120.

HSRC (Homeland Security Research Corporation). 2006. U.S. Bio-Detection Homeland Security Technology and Market Forecast: 2007-2012. Homeland Security Research [online]. Available: http://www.homelandsecurityresearch.com/uploads/HSRC-200606-T.pdf [accessed Mar. 11, 2008].

Hubbs, A.F., N.S. Minhas, W. Jones, M. Greskevitch, L.A. Battelli, D.W. Porter, W.T. Goldsmith, D. Frazer, D.P. Landsittel, J.Y. Ma, M. Barger, K. Hill, D. Schwegler-Berry, V.A. Robinson, and V. Castranova. 2001. Comparative pulmonary toxicity of 6 abrasive blasting agents. Toxicol. Sci. 61(1):135-143.

Hubbs, A.F., L.A. Batelli, W.T. Goldsmith, D.W. Porter, D. Frazer, S. Friend, D. Schwegler-Berry, R.R. Mercer, J.S. Reynolds, A. Grote, V. Castranova, G. Kullman, J.S. Fedan, J. Dowdy, and W.G. Jones. 2002. Necrosis of nasal and airway epithelium in rats inhaling vapors of artificial butter flavoring. Toxicol. Appl. Pharmacol. 185(2):128-135.

Hubbs, A.F., M. Greskevitch, E. Kuempel, F. Suarez, and M. Toraason. 2005. Abrasive blasting agents: Designing studies to evaluate relative risk. J. Toxicol. Environ. Health A 68(11-12):999-1016.

IARC (International Agency for Research on Cancer). 1989. Diesel and Gasoline Engine Exhausts and Some Nitroarenes. IARC Monographs on the Evaluation of Carcinogenic Risk of Chemicals to Humans, Vol. 46. Lyon, France: International Agency for Research on Cancer. 458 pp.

IARC (International Agency for Research on Cancer). 1993. Beryllium, Cadmium, Mercury, and Exposures in the Glass Manufacturing Industry. IARC Monographs on the Evaluation of Carcinogenic Risks to Humans, Vol. 58. Lyon, France: International Agency for Research on Cancer. 444 pp.

IARC (International Agency for Research on Cancer). 1997. Silica, Some Silicates, Coal Dust and Para-aramid Fibrils. IARC Monograph on the Evaluation of the Carcinogenic Risk of Chemicals to Humans Vol. 68. Lyon: International Agency for Research on Cancer. 506 pp.

IOM (Institute of Medicine). 2001. Tuberculosis in the Workplace. Washington, DC: National Academy Press.

IOM (Institute of Medicine). 2004. Damp Indoor Spaces and Health. Washington, DC: The National Academies Press.

IOM (Institute of Medicine). 2006. Reusability of Facemasks during an Influenza Pandemic: Facing the Flu. Washington DC: The National Academies Press.

Izbicki, G., R. Chavko, G.I. Banauch, M.D. Weiden, K.I. Berger, T.K. Aldrich, C. Hall, K.J. Kelly, and D.J. Prezant. 2007. World Trade Center "sarcoid-like" granulomatous pulmonary disease in New York City Fire Department rescue workers. Chest 131(5):1414-1423.

Johnson, B., D.R. Winters, T.R. Shreeve, and C.C. Coffey. 1998. Respirator filter reuse test using the laboratory simulant *Mycobacterium tuberculosis* (H37Ra STRAIN). J. Am. Biol. Safety Assoc. 3(3):105-116.

Kanwal, R., G. Kullman, C. Piacitelli, R. Boylstein, N. Sahakian, S. Martin, K. Fedan, and K. Kreiss. 2006. Evaluation of flavorings-related lung disease risk at six microwave popcorn plants. J. Occup. Environ. Med. 48(2):149-157.

Kelsey, K.T. 2007. Genetics and occupational safety and health. Occup. Environ. Med. 64(11):720-721.

Klink, K.J., and B.J. Meade. 2003. Dermal exposure to 3-amino-5-mercapto-1,2,4-triazole (AMT) induces sensitization and airway reactivity in BALB/c mice. Toxicol. Sci. 75(1):89-98.

Kreiss, K., and J. Cox-Ganser. 1997. Metalworking fluid-associated hypersensitivity pneumonitis: A workshop summary. Am. J. Ind. Med. 32(4):423-432.

Kreiss, K., A. Gomaa, G. Kullman, K. Fedan, E.J. Simoes, and P.L. Enright. 2002. Clinical bronchiolitis obliterans in workers at a microwave-popcorn plant. N. Engl. J. Med. 347(5):330-338.

KTA-Tator, Inc. 1998. Evaluation of Substitute Materials for Silica Sand in Abrasive Blasting. Contract No. 200-95-2946. Prepared for U.S. Department of Health and Human Services, Centers for Disease Control and Prevention, National Institute for Occupational Safety and Health, by KTA-Tator, Inc., Pittsburgh, PA. September 1998 [online]. Available: http://www.cdc.gov/niosh/pdfs/ab_p1rep.pdf [accessed Aug. 15, 2007].

Kuempel, E.D., L.T. Stayner, M.D. Attfield, and C.R. Buncher. 1995. Exposure-response analysis of mortality among coal miners in the U.S. Am. J. Ind. Med. 28(2):167-184.

Kuempel, E.D., C.L. Tran, A.J. Bailer, D.W. Porter, A.F. Hubbs, and V. Castranova. 2001. Biological and statistical approaches to predicting human lung cancer risk from silica. J. Environ. Pathol. Toxicol. Oncol. 20(Suppl. 2):15-32.

Lambert, G.O., J.R. Spurzem, D.J. Romberger, T.A. Wyatt, E. Lyden, A.M. Stromquist, J.A. Merchant, and S. Von Essen. 2005. Tumor necrosis factor-alpha hyper-responsiveness to endotoxin in whole blood is associated with chronic bronchitis in farmers. J. Agromed. 10(1):39-44.

Lee, S.A., S.A. Grinshpun and T. Reponen. 2005. Efficiency of N95 Filtering Facepiece Respirators and Surgical Masks Against Airborne Particles of Viral Size Range: Tests with Human Subjects. Presentation at the American Industrial Hygiene Conference and Expo, May 21-26, 2005, Anaheim, CA.

Leigh, J.P., P.S. Romano, M.B. Schenker, and K. Kreiss. 2002. Costs of occupational COPD and asthma. Chest 121(1):264-272.

Lemasters, G.K., J.E. Lockey, L.S. Levin, R.T. McKay, C.H. Rice, E.P. Horvath, D.M. Papes, J.W. Lu, and D.J. Feldman. 1998. An industry-wide pulmonary study of men and women manufacturing refractory ceramic fibers. Am. J. Epidemiol. 148(9):910-919.

Lockey, J.E., L.S. Levin, G.K. Lemasters, R.T. McKay, C.H. Rice, K.R. Hansen, D.M. Papes, S. Simpson, and M. Medvedovic. 1998. Longitudinal estimates of pulmonary function in refractory ceramic fiber manufacturing workers. Am. J. Respir. Crit. Care Med. 157(4 Pt 1):1226-1233.

Martin, S.B., Jr., and C.C. Coffey. 2005. NIOSH Health Hazard Evaluation Report: Salvation Army Harbor Light Center, St. Louis, Missouri. HETA 2003-0346-2969. U.S. Department of Health and Human Services, Centers for Disease Control and Prevention, National Institute for Occupational Safety and Health, Cincinnati, OH [online]. Available: http://www.cdc.gov/niosh/hhe/reports/pdfs/2003-0346-2969.pdf [accessed Aug. 17, 2007].

Maxfield, A.M., M.J. Lewis, J.A. Tisdale, S. Lachenmayr, and M. Lum. 1999. Effects of a preventive message in the organizational context: Occupational latex allergy in hospitals. Am. J. Ind. Med. (Suppl. 1):125-127.

Maxfield, A.M., J. Lewis, S. Lachenmayr, J. Tisdale, and M. Lum. 2000. A National Institute for Occupational Safety and Health Alert sent to hospitals and the intentions of hospital decision makers to advocate for latex allergy control measures. Health Educ. Res. 15(4):463-467.

Maynard, A.M., and E.D. Kuempel. 2005. Airborne nanostructured particles and occupational health. J. Nanoparticle Res. 7(6):587-614.

McConnell, R.S., and R.W. Hartle. 1986. Health Hazard Evaluation Report: International Bakers Service, Inc., South Bend, Indiana, HETA 85-171-1710. U.S. Department of Health and Human Services, Centers for Disease Control, National Institute for Occupation Safety and Health, Cincinnati, OH. July 1986 [online]. Available: http://www.defendingscience.org/case_studies/upload/NIOSH_1985_Intl_Bakers.pdf [accessed Aug. 16, 2007].

Mendell, M.J., G.M. Naco, T.G. Wilcox, and W.K. Sieber. 2003. Environmental risk factors and work-related lower respiratory symptoms in 80 office buildings: An exploratory analysis of NIOSH data. Am. J. Ind. Med. 43(6):630-641.

Menzies, D., and K. Kriess. 2006. Building-related illnesses. Pp. 737-783 in Asthma in the Workplace, 3rd Ed., I.L. Bernstein, M. Chan-Yeung, J.L. Malo, and D.I. Bernstein, eds. New York: Taylor & Francis.

Merchant, J.A., and C. Ortmeyer. 1981. Mortality of employees of two cotton mills in North Carolina. Chest 79(Suppl. 4):S6-S11.

Metalworking Fluids Standards Advisory Committee. 1999. Final Report of the OSHA Metalworking Fluids Standards Advisory Committee. July 15, 1999 [online]. Available http://www.osha.gov/SLTC/metalworkingfluids/mwf_finalreport.html [accessed Nov. 26, 2007

Miller, S.L., M. Hernandez, K. Fennelly, J. Martyny, and J. Macher. 2002. Efficacy of Ultraviolet Irradiation in Controlling the Spread of Tuberculosis. Prepared for Centers for Disease Control and Prevention, National Institute for Occupational Safety and Health [online]. Available: http://www.cdc.gov/niosh/reports/contract/pdfs/ultrairrTB.pdf [accessed Aug. 17, 2007].

MMWR (Morbidity and Mortality Weekly Report). 1987. Epidemiologic notes and reports occupational asthma from inhaled egg protein-Iowa. MMWR 36(2):23-25.

MMWR (Morbidity and Mortality Weekly Report). 1989a. A strategic program for the elimination of tuberculosis in the United States. MMWR 38(S-3):1-25.

MMWR (Morbidity and Mortality Weekly Report). 1989b. Publication of NIOSH Current Intelligence Bulletin on Carcinogenic Effects of Diesel Exhaust. MMWR 38(5):76-77.

MMWR (Morbidity and Mortality Weekly Report). 1990. Current trends silicosis: Cluster in sandblasters—Texas, and occupational surveillance for silicosis. MMWR 39(25):433-437.

MMWR (Morbidity and Mortality Weekly Report). 1992. National action plan to combat multidrug-resistant tuberculosis. MMWR 41(RR-11):1:48.

MMWR (Morbidity and Mortality Weekly Report). 1997. Chronic interstitial lung disease in nylon flocking industry workers—Rhode Island, 1992-1996. MMWR 46(38):897-901.

MMWR (Morbidity and Mortality Weekly Report). 2004. Silicosis in dental laboratory technicians—five states, 1994-2000. MMWR 53(9):195-197.

MMWR (Morbidity and Mortality Weekly Report). 2005. Guidelines for preventing the transmission of *Mycobacterium tuberculosis* in health-care settings, 2005. MMWR 54(RR-17):1-147.

MMWR (Morbidity and Mortality Weekly Report). 2006a. Advanced cases of coal workers' pneumoconiosis—two counties, Virginia, 2006. MMWR 55(33):909-913.

MMWR (Morbidity and Mortality Weekly Report). 2006b. Prevention and control of tuberculosis in correctional and detention facilities: Recommendations from CDC. MMWR 55(RR-9):1-44.

MMWR (Morbidity and Mortality Weekly Report). 2007a. Advanced pneumoconiosis among working underground coal miners—eastern Kentucky and southwestern Virginia, 2006. MMWR 56(26):652-655.

MMWR (Morbidity and Mortality Weekly Report). 2007b. Fixed obstructive lung disease among workers in the flavor-manufacturing industry—California, 2004-2007. MMWR 56(16):389-393.

Monforton, C. 2006. Weight of the evidence or wait for the evidence? Protecting underground miners from diesel particulate matter. Am. J. Public Health 96(2):271-276.

NAS (National Academies of Sciences). 2005. Framework for Review of Research Programs of the National Institute of Occupational Safety and Health. National Academies of Sciences, Washington, DC. December 12, 2005 [online]. Available: http://www.cdc.gov/niosh/nas/pdfs/Framework121905.pdf [accessed Aug. 17, 2007].

Nel, A., T. Xia, L. Madler, and N. Li. 2006. Toxic potential of materials at the nanolevel. Science 311(5761):622-627.

Newman, L.S., M.M. Mroz, R. Balkissoon, and L.A. Maier. 2005. Beryllium sensitization progresses to chronic beryllium disease: A longitudinal study of disease risk. Am. J. Respir. Crit. Care Med. 171(1):54-60.

NIH (National Institutes of Health). 2006. Prevention of Airborne Infections in Occupational Settings. NIH Guide for Grants and Contracts. Request for Applications (RFA) Number RFA-OH-06-002. U.S. Department of Health and Human Services, Centers for Disease Control, National Institute for Occupational Safety and Health [online]. Available: http://grants2.nih.gov/grants/guide/rfa-files/RFA-OH-06-002.html [accessed Aug. 22, 2007].

NIOSH (National Institute of Occupational Safety and Health). 1972. Criteria for a Recommended Standard: Occupational Exposure to Asbestos. NIOSH Publication No. 72-10267. U.S. Department of Health, Education, and Welfare, Public Health Service, Center for Disease Control, National Institute of Occupational Safety and Health [online]. Available: http://www.cdc.gov/niosh/docs/72-10267/ [accessed Aug. 13, 2007].

NIOSH (National Institute of Occupational Safety and Health). 1974. Pp. 54-55, 60-61 in Criteria for a Recommended Standard: Occupational Exposure to Crystalline Silica. DHHS (NIOSH) Publication No. 75-120. U.S. Department of Health, Education, and Welfare, Public Health Service, Center for Disease Control, National Institute of Occupational Safety and Health [online]. Available: http://www.cdc.gov/Niosh/75-120.html [accessed Aug. 15, 2007].

NIOSH (National Institute of Occupational Safety and Health). 1975. Criteria for a Recommended Standard: Occupational Exposure to Chromium (VI). HEW (NIOSH) Publication No. 76-129. U.S. Department of Health, Education, and Welfare, Public Health Service, Center for Disease Control, National Institute of Occupational Safety and Health [online]. Available: http://www.cdc.gov/niosh/pdfs/76-129a.pdf [accessed Aug. 28, 2007].

NIOSH (National Institute for Occupational Safety and Health). 1986a. Proposed National Strategy for the Prevention of Occupational Lung Disease. DHHS Publication No. 89-128. U.S. Department of Health and Human Services, Public Health Service, Centers for Disease Control, National Institute for Occupational Safety and Health, Cincinnati, OH. 17pp [online]. Available: http://www.cdc.gov/niosh/docs/89-128/pdfs/89-128.pdf [accessed Aug. 14, 2007].

NIOSH (National Institute of Occupational Safety and Health). 1986b. Occupational Respiratory Diseases. DHHS (NIOSH) Publication No. 86-102. U.S. Department of Health and Human Services, Public Health Service, Centers for Disease Control, National Institute for Occupational Safety and Health, Cincinnati, OH [online]. Available: http://www.cdc.gov/niosh/86-102.html [accessed Aug. 15, 2007].

NIOSH (National Institute of Occupational Safety and Health). 1992a. NIOSH Alert: Request for Assistance in Preventing Silicosis and Deaths from Sandblasting. DHHS (NIOSH) Publication No 92-102. U.S. Department of Health and Human Services, Public Health Service, Centers for Disease Control and Prevention, National Institute for Occupational Safety and Health, Cincinnati, OH. August 1992 [online]. Available: http://www.cdc.gov/niosh/92-102.html [accessed Aug. 15, 2007].

NIOSH (National Institute of Occupational Safety and Health). 1992b. NIOSH Alert: Request for Assistance in Preventing Silicosis and Deaths in Rock Drillers. DHHS (NIOSH) Publication No 92-107. U.S. Department of Health and Human Services, Public Health Service, Centers for Disease Control, National Institute for Occupational Safety and Health, Cincinnati, OH. August 1992 [online]. Available: http://www.cdc.gov/niosh/92-107.html [accessed Aug. 16, 2007].

NIOSH (National Institute of Occupational Safety and Health). 1994. Hazard Valuation and Technical Assistance Report: Jim Walter Resources, Inc., Brookwood, AL, Morgantown, WV. NIOSH Report No. HETA 94-0027. U.S. Department of Health and Human Services, Public Health Service, Centers for Disease Control and Prevention, National Institute for Occupational Safety and Health, Cincinnati, OH.

NIOSH (National Institute of Occupational Safety and Health). 1995. Criteria for a Recommended Standard: Occupation Exposure to Respirable Coal Mine Dust. DHHS (NIOSH) Publication No 95-106. U.S. Department of Health and Human Services, Public Health Service, Centers for Disease Control, National Institute for Occupational Safety and Health, Cincinnati, OH. September 1995 [online]. Available: http://www.cdc.gov/niosh/95-106.html [accessed Aug. 16, 2007].

NIOSH (National Institute for Occupational Safety and Health). 1996a. National Occupational Research Agenda. DHHS Publication No. 96-115. U.S. Department of Health and Human Services, Public Health Service, Centers for Disease Control, National Institute for Occupational Safety and Health, Cincinnati, OH [online]. Available: http://www.cdc.gov/niosh/docs/96-115/ [accessed March 3, 2007].

NIOSH (National Institute for Occupational Safety and Health). 1996b. Preventing Asthma and Death from Diisocyanate. DHHS Publication No. 96-111. U.S. Department of Health and Human Services, Public Health Service, Centers for Disease Control and Prevention, National Institute for Occupational Safety and Health, Cincinnati, OH [online]. Available: http://www.cdc.gov/niosh/asthma.html [accessed July 26, 2007].

NIOSH (National Institute for Occupational Safety Health). 1997a. Preventing Allergic Reactions to Natural Rubber Latex in the Workplace. DHHS Publication No. 97-135. U.S. Department of Health and Human Services, Public Health Service, Centers for Disease Control and Prevention, National Institute for Occupational Safety and Health, Cincinnati, OH [online]. Available: http://www.cdc.gov/niosh/latexalt.html [accessed June 29, 2007].

NIOSH (National Institute for Occupational Safety Health). 1997b. Cooperative Agreement for State-Based Surveillance Activities—Sentinel Event Notification Systems for Occupational Risk (SENSOR); Notice of Availability of Funds for Fiscal Year 1998. Billing Code: 4163-19-P, Program Announcement 708. U.S. Department of Health and Human Services, Centers for Disease Control and Prevention, National Institute for Occupational Safety and Health [online]. Available: http://www.cdc.gov/niosh/oep/frn708.html [accessed Aug. 17, 2007].

NIOSH (National Institute for Occupational Safety Health). 1998a. Criteria for Recommended Standards: Occupational Exposure to Metalworking Fluids. DHHS (NIOSH) Publication No. 98-102. U.S. Department of Health and Human Services, Public Health Service, Centers for Disease Control and Prevention, National Institute for Occupational Safety and Health, Cincinnati, OH [online]. Available: http://www.cdc.gov/niosh/pdfs/98-102-a.pdf [accessed Nov. 26, 2007].

NIOSH (National Institute for Occupational Safety Health). 1998b. Mycobacterium Tuberculosis, Airborne. Method: 0900, Issue 1: 15 January 1998. In: NIOSH Manual of Analytical Method, 4th Ed.[online]. Available: http://www.cdc.gov/niosh/nmam/pdfs/0900.pdf [accessed Aug. 16, 2007].

NIOSH (National Institute for Occupational Safety Health). 1998c. Testimony of the National Institute for Occupational Safety and Health on the Occupational and Health Administration Proposed Rule on Occupational Exposure to Tuberculosis; 29 CFR Part 1910.1035, Docket No. H-371. Presentation at the OSHA Informal Public Hearing, April 17, 1998, Washington, DC.

NIOSH (National Institute of Occupational Safety and Health). 2002a. NIOSH Hazard Review: Health Effects of Occupational Exposure to Respirable Crystalline Silica. DHHS (NIOSH) Publication No 2002-129. U.S. Department of Health and Human Services, Centers for Disease Control and Prevention, National Institute for Occupational Safety and Health [online]. Available: http://www.cdc.gov/niosh/02-129A.html [accessed Aug. 16, 2007].

NIOSH (National Institute of Occupational Safety and Health). 2002b. Guidance for Protecting Building Environments from Airborne Chemical, Biological and Radiological Attacks. DHHS (NIOSH) Publication No. 2002-139. U.S. Department of Health and Human Services, Centers for Disease Control and Prevention, National Institute for Occupational Safety and Health, Cincinnati, OH [online]. Available: http://www.cdc.gov/niosh/docs/2002-139/pdfs/2002-139.pdf [accessed Aug. 17, 2007].

NIOSH (National Institute for Occupational Safety and Health). 2002c. Protecting Workers from Anthrax Infection—Response from the National Institute for Occupational Safety and Health. DHHS (NIOSH) Publication No 2002-142. U.S. Department of Health and Human Services, Public Health Service, Centers for Disease Control and Prevention, National Institute for Occupational Safety and Health, Cincinnati, OH [online]. Available: http://0-www.cdc.gov.mill1.sjlibrary.org/niosh/pdfs/02-142.pdf [accessed Aug. 17, 2007].

NIOSH (National Institute for Occupational Safety and Health). 2003. Work-Related Lung Disease Surveillance Report: 2002. U.S. Department of Health and Human Services, Public Health Service, Centers for Disease Control and Prevention, National Institute for Occupational Safety and Health, Cincinnati, OH [online]. Available: http://www.cdc.gov/niosh/docs/2003-111/pdfs/2003-111.pdf [accessed February 7, 2007].

NIOSH (National Institutes for Occupational Safety and Health). 2004. Worker Health Chartbook 2004. NIOSH Publication No. 2004-146. U.S. Department of Health and Human Services, Public Health Service, Centers for Disease Control and Prevention, National Institute for Occupational Safety and Health, Cincinnati, OH [online]. Available: http://www.cdc.gov/niosh/docs/chartbook/ [accessed Aug. 27, 2007].

NIOSH (National Institute of Occupational Safety and Health). 2005a. Comments of the National Institute for Occupational Safety and Health on the Mine Safety and Health Administration Proposed Rule on Asbestos Exposure Limit. U.S. Department of Health and Human Services, Public Health Service, Centers for Disease Control and Prevention, National Institute for Occupational Safety and Health. October 13, 2005 [online]. Available: http://www.cdc.gov/niosh/review/public/099/pdfs/Asbestos-msha_final%202005_proposed%20rule.pdf [accessed Aug. 13, 2007].

NIOSH (National Institute of Occupational Safety and Health). 2005b. Strategic Plan for NIOSH Nanotechnology Research Filling the Knowledge Gaps. Nanotechnology Research Program, Centers for Disease Control and Prevention, National Institute for Occupational Safety and Health. September 28, 2005.

NIOSH (National Institute for Occupational Safety and Health). 2006a. Respiratory Disease Research Program, National Academy Review 2006.

NIOSH (National Institute for Occupational Safety and Health). 2006b. Respiratory Disease Research Program Evidence Package. Materials submitted for Committee.

NIOSH (National Institute for Occupational Safety and Health). 2006c. Preventing Asthma and Death from MDI Exposure during Spray-on Truck Bed Liner and Related Applications. DHHS Publication No. 2006-149. U.S. Department of Health and Human Services, Centers for Disease Control and Prevention, National Institute for Occupational Safety and Health, Cincinnati, OH [online]. Available: http://www.cdc.gov/niosh/docs/2006-149/default.html [accessed June 29, 2007].

NIOSH (National Institute of Occupation Safety and Health). 2006d. National Occupational Respiratory Mortality System (NORMS). U.S. Department of Health and Human Services, Public Health Service, Centers for Disease Control and Prevention, National Institute for Occupational Safety and Health, Division of Respiratory Disease Studies, Surveillance Branch [online]. Available: http://webappa.cdc.gov/ords/norms.html [accessed August 10, 2007].

NIOSH (National Institute of Occupation Safety and Health). 2006e. Criteria for a Recommended Standard: Occupational Exposure to Refractory Ceramic Fibers. DHHS (NIOSH) Publication No. 2006-123. U.S. Department of Health and Human Services, Public Health Service, Centers for Disease Control and Prevention, National Institute for Occupational Safety and Health, Cincinnati, OH [online]. Available: http://www.cdc.gov/niosh/docs/2006-123/ [accessed Aug. 16, 2007].

NIOSH (National Institute for Occupational Safety Health). 2006f. Approaches to Safe Nanotechnology: An Information Exchange with NIOSH. U.S. Department of Health and Human Services, Centers for Disease Control and Prevention, National Institute for Occupational Safety and Health [online]. Available: http://www.cdc.gov/niosh/topics/nanotech/safenano/pdfs/approaches_to_safe_nanotechnology_28november2006_updated.pdf [accessed Aug. 27, 2007].

NIOSH (National Institute of Occupation Safety and Health). 2007a. World-Related Lung Disease (eWoRLD) Surveillance System. U.S. Department of Health and Human Services, Centers for Disease Control and Prevention, National Institute for Occupational Safety and Health [online]. Available: http://www2.cdc.gov/drds/WorldReportData/ [accessed Nov. 16, 2007].

NIOSH (National Institute of Occupation Safety and Health). 2007b. Understanding Respiratory Protection Against SARS [online]. Available: http://www.cdc.gov/niosh/npptl/topics/respirators/factsheets/respsars.html [accessed June 28, 2007].

NIOSH (National Institute for Occupational Safety Health). 2007c. Avian Influenza: Protecting Workers from Exposure. NIOSH Safety and Health Topic. U.S. Department of Health and Human Services, Centers for Disease and Prevention, National Institute for Occupational Safety and Health [online]. Available: http://www.cdc.gov/niosh/topics/avianflu/ [accessed Aug 27, 2007].

NIOSH (National Institute for Occupational Safety Health). 2007d. Worker Notification Program. [online]. Available: http://www.cdc.gov/niosh/pgms/worknotify/ [accessed May 22, 2007].

NIOSH (National Institute for Occupational Safety Health). 2007e. Progress toward Safe Nanotechnology in the Workplace. DHHS Publication No. 2007–123. U.S. Department of Health and Human Services, Centers for Disease and Prevention, National Institute for Occupational Safety and Health [online]. Available: http://www.cdc.gov/niosh/docs/2007-123/pdfs/2007-123.pdf [accessed July 26, 2007].

Noll, J.D., R.J. Timko, L.J. McWilliams, P. Hall, and R. Haney. 2005. Sampling results of the improved SKC diesel particulate matter cassette. J. Occup. Environ. Hyg. 2(1):29-37.

NTP (National Toxicology Program). 1998. Report on Carcinogens, 8th Ed. U.S. Department of Health and Human Service, Public Health Service, National Toxicology Program [online]. Available: http://ntp.niehs.nih.gov/index.cfm?objectid=06F2A7E4-0B2E-B281-377642BF50C8A936 [accessed July 26, 2007].

NTP (National Toxicology Program). 2005. Report on Carcinogens, 11th Ed. U.S. Department of Health and Human Service, Public Health Service, National Toxicology Program [online]. Available: http://ntp.niehs.nih.gov/index.cfm?objectid=32BA9724-F1F6-975E-7FCE50709CB4C932 [accessed July 26, 2007].

Oberdorster, G., A. Maynard, K. Donaldson, V. Castranova, J. Fitzpatrick, K. Ausman, J. Carter, B. Karn, W. Kreyling, D. Lai, S. Olin, N. Monterio-Riviere, D. Warheit, and H. Yang. 2005. Principles for characterizing the potential human effects from exposure to nanomaterials: Elements of screening strategy. Part. Fibre Toxicol. 2:8.

Ortega, H.G., D.N. Weissman, D.L. Carter, and D. Banks. 2002. Use of specific inhalation challenge in the evaluation of workers at risk for occupational asthma: A survey of pulmonary, allergy, and occupational medicine residency training programs in the U.S. and Canada. Chest 121(4):1323-1328.

Palmer, K.T., J. Poole, R.G. Rawbone, and D. Coggon. 2004. Quantifying the advantages and disadvantages of pre-placement genetic screening, Occup. Environ. Med. 61(5):448-453.

Park, R.M., and L.T. Stayner. 2006. A search for thresholds and other nonlinearities in the relationship between hexavalent chromium and lung cancer. Risk Anal. 26(1):79-88.

Park, R.M., J.F. Bena, L.T. Stayner, R.J. Smith, H.J. Gibb, and P.S. Lees. 2004. Hexavalent chromium and lung cancer in the chromate industry: A quantitative risk assessment. Risk Anal. 24(5):1099-1108.

Park, R.M., R.M. Bowler, D.E. Eggerth, E. Diamond, K.J. Spencer, D. Smith, and R. Gwiazda. 2006. Issues in neurological risk assessment for occupational exposures: The Bay Bridge welders. Neurotoxicology 27(3):373-384.

Pechter, E., L.K. Davis, C. Tumpowsky, J. Flattery, R. Harrison, F. Reinisch, M.J. Reilly, K.D. Rosenman, D.P. Schill, D. Valiante, and M. Filios. 2005. Work-related asthma among health care workers: Surveillance data from California, Massachusetts, Michigan, and New Jersey, 1993-1997. Am. J. Ind. Med. 47(3):265-275.

Perry, W. 2006. Letter to D.N. Weissman, Director, Division of Respiratory Disease Studies, NIOSH, Morgantown, WV, from W. Perry, U.S. Department of Labor, Occupational Safety and Health Administration, Washington, DC. June 1, 2006.

Piacitelli, C., and V. Antao. 2006. Health Hazard Evaluation Report: Hallmark Cards, Inc., Lawrence, Kansas. HETA-2004-0013-2990. U.S. Department of Health and Human Services, Centers for Disease Control and Prevention, National Institute for Occupational Safety and Health. January 2006 [online]. Available: http://www.cdc.gov/niosh/hhe/reports/pdfs/2004-0013-2990.pdf [accessed Aug. 16, 2007].

Piacitelli, G.M., W.K. Sieber, D.M. O'Brien, R. Hughes, R.A. Glaser, and J.D. Catalano. 2001. Metalworking fluid exposures in small machine shops: An overview. Am. Ind. Hyg. Assoc. J. 62(3):356-370.

Porter, D.W., V. Castranova, V.A. Robinson, A.F. Hubbs, R.R. Mercer, J. Scabilloni, T. Goldsmith, D. Schwegler-Berry, L. Battelli, R. Washko, J. Burkhart, C. Piacitelli, M. Whitmer, and W. Jones. 1999. Acute inflammatory reaction in rats after intratracheal instillation of material collected from a nylon flocking plant. J. Toxicol. Environ. Health A 57(1):25-45.

Porter, D.W., A.F. Hubbs, V.A. Robinson, L.A. Battelli, M. Greskevitch, M. Barger, D. Landsittel, W. Jones, and V. Castranova. 2002. Comparative pulmonary toxicity of blasting sand and five substitute abrasive blasting agents. J. Toxicol. Environ. Health A 65(16):1121-1140.

Porter, D.W., S.S. Leonard, V. Castranova. 2006. Particles and cellular oxidative and nitrosative stress. Pp. 119-138 in Particle Toxicology, K. Donaldson, and P. Borm, eds. Boca Raton: CRC Press.

Qian, Y., K. Willeke, S.A. Grinshpun, and J. Donnelly. 1997. Performance of N95 respirators: Reaerosolizaton of bacteria and solid particles. Am. Ind. Hyg. Assoc. J. 58(12):876-880.

Qian, Y., K. Willeke, S.A. Grinshpun, J. Donnelly, and C.C. Coffey. 1998. Performance of N95 respirators: Filtration efficiency for airborne microbial and inert particles. Am. Ind. Hyg. Assoc. J. 59(2):128-132.

Rosenman, K.D., M.J. Reilly, and P.K. Henneberger. 2003. Estimating the total number of newly-recognized silicosis cases in the U.S. Am. J. Ind. Med. 44(2):141-147.

Rosenman, K., V. Hertzberg, C. Rice, M.J. Reilly, J. Aronchick, J.E. Parker, J. Regovich, and M. Rossman. 2005. Chronic beryllium disease and sensitization at a beryllium processing facility. Environ. Health Perspect. 113(10):1366-1372.

Sackett, H.M., L.A. Maier, L.J. Silveira, M.M. Mroz, L.G. Ogden, J.R. Murphy, and L.S. Newman. 2004. Beryllium medical surveillance at a former nuclear weapons facility during cleanup operations. J. Occup. Environ. Med. 46(9):953-961.

Sama, S.R., D.K. Milton, P.R. Hunt, E.A. Houseman, P.K. Henneberger, and R.A. Rosiello. 2006. Case-by-case assessment of adult-onset asthma attributable to occupational exposures among members of a health maintenance organization. J. Occup. Environ. Med. 48(4):400-407.

Sargent, L.M., J.R. Senft, D.T. Lowry, A.M. Jefferson, F.L. Tyson, A.M. Malkinson, A.E. Coleman, and S.H. Reynolds. 2002. Specific chromosomal aberrations in mouse lung adenocarcinoma cell lines detected by spectral karyotyping: A comparison with human lung adenocarcinoma. Cancer Res. 62(4):1152-1157.

Schnakenberg, Jr., G.H., and A.D. Bugarski. 2002. Review of Technology Available to the Underground Mining Industry for Control of Diesel Emissions. Information Circular 9462. U.S. Department of Health and Human Services, Public Health Service, Centers for Disease Control and Prevention, National Institute for Occupational Safety and Health, Pittsburgh Research Laboratory, Pittsburgh, PA [online]. Available: http://www.cdc.gov/niosh/mining/pubs/pdfs/ic9462.pdf [accessed Aug. 21, 2007].

Schnakenberg, Jr., G.H., and A.D. Bugarski. 2004. Testing diesel emission control techniques. Proceedings of the Diesel Particulate in Mining Conference, June 8-9, 2004, Yeppoon, Queensland, Australia.

Schnakenberg, Jr., G.H., A.D. Bugarski, J. Angel, and G. Saseen. 2005. Metal and Nonmetal Diesel Particulate Filter Selection Guide. U.S. Department of Labor, Mine Safety and Health Administration [online]. Available: http://www.msha.gov/nioshmnmfilterselectionguide/dpmfilterguide.htm [accessed June 27, 2007].

Schuler, C.R., M.S. Kent, D.C. Deubner, M.T. Berakis, M. McCawley, P.K. Henneberger, M.D. Rossman, and K. Kreiss. 2005. Process-related risk of beryllium sensitization and disease in a copper-beryllium alloy facility. Am. J. Ind. Med. 47(3):195-205.

Schulte, P.A. 2007. The contributions of genetics and genomics to occupational safety and health. Occup. Environ. Med. 64(11):717-718.

Sepkowitz, K.A., and L. Eisenberg. 2005. Occupational deaths among healthcare workers. Emerg. Infect. Dis. 11(7):1003-1008.

Shi, X., V. Castranova, B. Halliwell, and V. Vallyathan. 1998. Reactive oxygen species and silica-induced carcinogenesis. J. Toxicol. Environ. Health B Crit. Rev. 1(3):181-197.

Shvedova, A.A., E.R. Kisin, R. Mercer, A.R. Murray, V.J. Johnson, A.I. Potapovich, Y.Y. Tyurina, O. Gorelik, S. Arepalli, D. Schwegler-Berry, A.F. Hubbs, J. Antonini, D.E. Evans, B.K. Ku, D. Ramsey, A. Maynard, V.E. Kagan, V. Castranova, and P. Baron. 2005. Unusual inflammatory and fibrogenic pulmonary responses to single-walled carbon nanotubes in mice. Am. J. Physiol. Lung Cell. Mol. Physiol. 289(5):L698-L708.

Simeonova, P.P., W. Toriumi, C. Kommineni, M. Erkan, A.E. Munson, W.N. Rom, and M.I. Luster. 1997. Molecular regulation of IL-6 activation by asbestos in lung epithelial cells: Role of reactive oxygen species. J. Immunol. 159(8):3921-3928.

Snyder, J.A., A. Weston, S.S. Tinkle, and E. Demchuk. 2003. Electrostatic potential on human leukocyte antigen: Implications for putative mechanism of chronic beryllium disease. Environ. Health Perspect. 111(15):1827-1834.

Stayner, L.T., D. Dankovic, R. Smith, and K. Steenland. 1998. Predicted lung cancer risk among miners exposed to diesel exhaust particles. Am. J. Ind. Med. 34(3):207-219.

Steenland, K. and D. Brown. 1995. Mortality study of gold miners exposed to silica and non-asbestiform amphibole minerals: An update with 14 more years of follow-up. Am. J. Ind. Med. 27(2):217-229.

Steenland, K., D. Loomis, C. Shy, and N. Simonsen. 1996. Review of occupational lung carcinogens. Am. J. Ind. Med. 29(5):474-490.

Steenland, K., J. Deddens, and L. Stayner. 1998. Diesel exhaust and lung cancer in the trucking industry: Exposure-response analysis and risk assessment. Am. J. Ind. Med. 34(3):220-228.

Steenland, K., C. Burnett, N. Lalich, E. Ward, and J. Hurrell. 2003. Dying for work. The magnitude of U.S. mortality from selected causes of death associated with occupation. Am. J. Ind. Med. 43(5):461-482.

Stefaniak, A.B., M.D. Hoover, R.M. Dickerson, E.J. Peterson, G.A. Day, P.N. Breysse, M.S. Kent, and R.C. Scripsick. 2003. Surface area of respirable beryllium metal, oxide, and copper alloy aerosols and implications for assessment of exposure risk of chronic beryllium disease. Am. Ind. Hyg. Assoc. J. 64(3):297-305.

Stefaniak, A.B., M.D. Hoover, G.A. Day, R.M. Dickerson, E.J. Peterson, M.S. Kent, C.R. Schuler, P.N. Breysse, and R.C. Scripsick. 2004. Characterization of physicochemical properties of beryllium aerosols associated with chronic beryllium disease. J. Environ. Monit. 6(6):523-532.

Teshale, E.H., J. Painter, G.A. Burr, P. Mead, S.V. Wright, L.F. Cseh, R. Zabrocki, R. Collins, K.A. Kelley, J.L. Hadler, and D.L. Swerdlow. 2002. Environmental sampling for spores of *Bacillus anthracis*. Emerg. Infect. Dis. 8(10):1083-1087.

Toraason, M., R. Albertini, S. Bayard, W, Bigbee, A. Blair, P. Boffetta, S. Bonassi, S. Chanock, D. Christiani, D. Eastmond, S. Hanash, C. Henry, F. Kadlubar, F. Mirer, D. Nebert, S. Rapport, K. Rest, N. Rothman, A. Ruder, R. Savage, P. Schulte, J. Siemiatycki, P. Shields, M. Smith, P. Tolbert, R. Vermeulen, P. Vineis, S. Wacholder, E. Ward, M. Waters, and A. Weston. 2004. Applying new biotechnologies to the study of occupational cancer—a workshop summary. Environ. Health Perspect. 112(4):413-416.

USDHS (U.S. Department of Homeland Security). 2007. Lab-in-a-Box. S & T Snapshorts 1(2), June 2007 [online]. Available: http://www.homelandsecurity.org/snapshots/newsletter/2007-06.htm [accessed Mar. 11, 2008].

USPS (U.S. Postal Service). 2002. U.S. Postal Service Emergency Preparedness Plan for Protecting Postal Employees and Postal Customers From Exposure to Biohazardous Material and for Ensuring Mail Security Against Bioterror Attacks. March 6, 2002 [online]. Available: http://www.usps.com/news/2002/epp/emerprepplan.pdf [accessed Aug. 17, 2007].

USPS (U.S. Postal Service). 2005. U.S. Postal Service Bio Detection System (BDS). May 25, 2005 [online]. Available: http://www.nemaweb.org/?1381 [accessed Aug. 22, 2007].

Vainio, H. 2007. Genetics and occupational health and safety. Occup. Environ. Med. 64(11):721-722.

Vallyathan, V., V. Castranova, D. Pack, S. Leonard, J. Shumaker, A.F. Hubbs, D.A. Shoemaker, D.M. Ramsey, J.R. Pretty, J.L. McLaurin, A. Khan, and A. Teass. 1995. Freshly fractured quartz inhalation leads to enhanced lung injury and inflammation: Potential role of free radicals. Am. J. Crit. Care. Med. 152(3):1003-1009.

Vallyathan, V., V. Castranova, and X. Shi. 2002. Oxygen/Nitrogen Radicals: Cell Injury and Disease. Boston: Kluwer.

Vallyathan, V., V. Castranova, and X. Shi. 2004. Oxygen/Nitrogen Radicals: Lung Injury and Disease. New York: Marcel Dekker.

Van Dyke, K., and V. Castranova. 1987. Cellular Chemiluminescence. Boca Raton: CRC Press.

Vollmer, W.M., M.A. Heumann, V.R. Breen, P.K. Henneberger, E.A. O'Connor, J.M. Villnave, E.A. Frazier, and A.S. Buist. 2005. Incidence of work-related asthma in members of a health maintenance organization. J. Occup. Environ. Med. 47(12):1292-1297.

Wang, J., and K. Ashley. 2004. Method for the Determination of Hexavalent Chromium Using Ultrasonication and Strong Anion Exchange Solid Phase Extraction. U.S. Patent 6,808,931 [online]. Available: http://www.patentstorm.us/patents/6808931-claims.html [accessed Aug. 21, 2007].

Wang, J., K. Ashley, E.R. Kennedy, and C. Neumeister. 1997. Determination of hexavalent chromium in industrial hygiene samples using ultrasonic extraction and flow injection analysis. Analyst 122(11):1307-1312.

Wang, M.L., Z.E. Wu, Q.G. Du, E.L. Petsonk, K.L. Peng, Y.D. Li, S.K. Li, G.H. Han, and M.D. Atffield. 2005. A prospective cohort study among new Chinese coal miners: The early pattern of lung function change. Occup. Environ. Med. 62(11):800-805.

Ward, E., A. Ruder, and C. Mills. 1993. Assessment of Cancer Priority Setting in IWSB. Working Draft NIOSH Report. September 3, 1993. 45 pp.

Ward, E.M., P.A. Schulte, S. Bayard, A. Blair, P. Brandt-Rauf, M.A. Butler, D. Dankovic, A.F. Hubbs, C. Jones, M. Karstadt, G.L. Kedderis, R. Melnick, C.A. Redlich, N. Rothman, R.E. Savage, M. Sprinker, M. Toraason, A. Weston, A.F. Olshan, P. Stewart, and S.H. Zahm. 2003. Priorities for development of research methods in occupational cancer. Environ. Health Perspect. 111(1):1-12.

Ward, K. 2007. NIOSH reports black lung rates double since '97. Charleston Gazette, September 14, 2007 [online]. Available: http://www.wvgazette.com/section/News/2007091331 [accessed Oct. 3, 2007].

Washko, R., J. Burkhart, and C. Piacitelli. 1998. Health Hazard Evaluation Report: Microfibres, Inc., Pawtucket, RI. HETA 96-0093-2685.U.S. Department of Health and Human Services, Centers for Disease Control, National Institute for Occupation Safety and Health, Cincinnati, OH. April 1998 [online]. Available: http://0-www.cdc.gov.mill1.sjlibrary.org/niosh/hhe/reports/pdfs/1996-0093-2685.pdf [accessed Nov. 15, 2007].

Washko, R.M., B. Day, J.E. Parker, R.M. Castellan, and K. Kreiss. 2000. Epidemiologic investigation of respiratory morbidity at a nylon flock plant. Am. J. Ind. Med. 38(6):628-638.

Weissman, D. 2006. Overview of the Respiratory Diseases Research Program (RDRP). Presentation at the First Meeting on Review the NIOSH Respiratory Disease Research Program, October 26-27, 2006, Washington, DC [online]. Available: http://www.cdc.gov/niosh/nas/pdfs/Weissman-Presentation-to-NA-10-26-06.pdf [accessed Aug. 14, 2007].

Welch, L., K. Ringen, E. Bingham, J. Dement, T. Takaro, W. McGowan, A. Chen, and P. Quinn. 2004. Screening for beryllium disease among construction trade workers at Department of Energy nuclear sites. Am. J. Ind. Med. 46(3):207-218.

Wernli, K.J., R.M. Ray, D.L. Gao, D.B. Thomas, and H. Checkoway. 2003. Cancer among women textile workers in Shanghai, China: overall incidence patterns, 1989-1998. Am. J. Ind. Med. 44(6):595-599.

Wernli, K.J., E.D. Fitzgibbons, R.M. Ray, D.L. Gao, W. Li, N.S. Seixas, J.E. Camp, G. Astrakianakis, Z. Feng, D.B. Thomas, and H. Checkoway. 2006. Occupational risk factors for esophageal and stomach cancers among female textile workers in Shanghai, China. Am. J. Epidemiol. 163(8):717-725.

Weston, A., J. Snyder, E.C. McCanlies, C.R. Schuler, M.E. Andrew, K. Kreiss, and E. Demchuk. 2005. Immunogenetic factors in beryllium sensitization and chronic beryllium disease. Mutat. Res. 592(1-2):68-78.

WHO (World Health Organization). 1996. Diesel Fuel and Exhaust Emissions. Environmental Health Criteria 171. Geneva: World Health Organization [online]. Available: http://www.inchem.org/documents/ehc/ehc/ehc171.htm [accessed Aug. 21, 2007].

Woskie, S.R., A. Kalil, D. Bello, and M.A. Virji. 2002. Exposures to quartz, diesel, dust, and welding fumes during heavy and highway construction AIHA J. 63(4):447-457.

Xu, P., J. Peccia, P. Fabian, J.W. Martyny, K.P. Fennelly, M. Hernandez, and S.L. Miller. 2003. Efficacy of ultraviolet germicidal irradiation of upper-room air in inactivating airborne bacterial spores and mycobacteria in full-scale studies. Atmos. Environ. 37(3):405-419.

Xu, P., E. Kujundzic, J. Peccia, M.P. Schaffer, G. Moss, M. Hernandez, and S.L. Miller. 2005. Impact of environmental factors on efficacy of upper-room air ultraviolet germicidal irradiation for inactivating Mycobacteria. Environ. Sci. Technol. 39(24):9656-9664.

Yang, H.M., J.M. Antonini, M.W. Barger, L. Butterworth, B.R. Roberts, J.K.Ma, V. Castranova, and J.Y. Ma. 2001. Diesel exhaust particles suppress macrophage function and slow the pulmonary clearance of *Listeria monocytogenes* in rats. Environ. Health Perspect. 109(5):515-521.

Yuan, B.Z., A.M. Jefferson, K.T. Baldwin, S.S. Thorgeirsson, N.C. Popescu, and S.H. Reynolds. 2004. DLC-1 operates as a tumor suppressor gene in human non-small cell lung carcinomas. Oncogene 23(7):1405-1411.

Yucesoy, B., V. Vallyathan, D.P. Landsittel, D.S. Sharp, A. Weston, G.R. Burleson, P.P. Simenova, M. McKinstry, and M.I. Luster. 2001. Association of tumor necrosis factor-alpha and interleukin-1 gene polymorphisms with silicosis. Toxicol. Appl. Pharmacol. 172(1):75-82.

Zeidler, P.C., and V. Castranova. 2004. Role of nitric oxide in pathologic responses of the lung to exposure to environmental/occupational agents. Redox Rep. 9(1):7-18.

Zeiss, C.R., A. Gomaa, F.M. Murphy, D.N. Weissman, M. Hodgson, D. Foster, S. Dejativongse, K. Colella, K. Kestenberg, V. Kurup, R. Bush, A.M. Chiu, K.J. Kelly, and J.N. Fink. 2003. Latex hypersensitivity in Department of Veterans Affairs health care workers: Glove use, symptoms, and sensitization. Ann. Allergy Asthma Immunol. 91(6):539-545.

Zhang, X.D., P.D. Siegel, and D.M. Lewis. 2002. Immunotoxicology of organic acid anhydrides (OAAs). Int. Immunopharmacol. 2(2-3):239-248.

Zhang, X.D., J.S. Fedan, D.M. Lewis, and P.D. Siegel. 2004. Asthma like biphasic airway responses in Brown Norway rats sensitized by dermal exposure to dry trimellitic anhydride powder. J Allergy Clin. Immunol. 113(2):320-326.

Zhang, X.D., M.E. Andrew, A.F. Hubbs, and P.D. Siegel. 2006. Airway responses in Brown Norway rats following inhalation sensitization and challenge with trimellitic anhydride. Toxicol. Sci. 94(2):322-329.

Appendixes

A

Framework for the Review of Research Programs of the National Institute for Occupational Safety and Health[*]

This is a document prepared by the National Academies' Committee for the Review of NIOSH Research Programs,[1] also referred to as the Framework Committee. This document is not a formal report of the National Academies—rather, it is a framework proposed for use by a number of National Academies committees that will be reviewing research in various research programs and health-outcomes programs. This version will be posted on the website of the National Academies and NIOSH for review. It is a working document that will be subject to change by the Framework Committee aimed at improving its relevance on the basis of responses received from evaluation committee members, NIOSH, stakeholders, and the general public before and during the course of the assessments conducted by independent evaluation committees of up to 15 research programs and health-outcomes programs.

---

*Version of 12/19/05.

[1]Members of the National Academies' Committee for the Review of NIOSH Research Programs include: David Wegman (Chair; University of Massachusetts Lowell School of Health and Environment), William Bunn, III (International Truck and Engine Corporation), Carlos Camargo (Harvard Medical School), Susan Cozzens (Georgia Institute of Technology), Letitia Davis (Massachusetts Department of Public Health), James Dearing (Kaiser Permanente), Fred Mettler, Jr. (University of New Mexico School of Medicine), Franklin Mirer (Hunter School of Health Sciences), Jacqueline Nowell (United Food and Commercial Workers International Union), Raja Ramani (Pennsylvania State University), Jorma Rantanen (Finnish Institute of Occupational Health), Rosemary Sokas (University of Illinois at Chicago School of Public Health), Richard Tucker (Tucker and Tucker Consultants, Inc. and University of Texas at Austin), and James Zuiches (North Carolina State University).

All public comments submitted to the Committee for the Review of NIOSH Research Programs will be included in the Public Access File for this study as provided in the National Academies Terms of Use (www.nationalacademies.org/legal/terms.html). Please keep in mind that if you directly disclose personal information in your written comments, this information may be collected and used by others.

CONTENTS

Acronyms used in this document
- I. Overview of Charge
  - I.A. NIOSH Strategic Goals and Operational Plan
  - I.B. Information from Other Evaluations
  - I.C. Evaluation Committees
  - I.D. Evaluation Committees' Information Needs
- II. Summary of Evaluation Process
  - II.A. The Evaluation Flow Chart (Figure 2)
  - II.B. Steps in Program Evaluation
  - II.C. Assessing Relevance
  - II.D. Assessing Impact
- III. Evaluation of NIOSH Research Programs—the Process
  - III.A. Analysis of External Factors Relevant to the NIOSH Program
    - III.A.1. Overview
    - III.A.2. Considerations for Discussion
  - III.B. Evaluating NIOSH Research Programs (Addressing Charges 1 and 2)
    - III.B.1. Identifying Period of Time to be Evaluated
    - III.B.2. Identification of Major Challenges (Circle in Figure 2)
    - III.B.3. Analysis of NIOSH Program Strategic Goals and Objectives (Box A in Figure 2)
    - III.B.4. Review of Inputs (Box B in Figure 2)
    - III.B.5. Review of Activities (Box C in Figure 2)
    - III.B.6. Review of Outputs (Box D in Figure 2)
    - III.B.7. Review of Intermediate Outcomes (Box E in Figure 2)
    - III.B.8. Review of End Outcomes (Box F in Figure 2)
    - III.B.9. Review of Other Outcomes
    - III.B.10. Summary Evaluation Ratings and Rationale
  - III.C. Identifying Significant Emerging Research (addressing Charge 3)
- IV. Evaluation Committee Report Template
- V. Framework Committee Final Report

Figure 1 The NIOSH operational plan presented as a logic model
Figure 2 Flow chart for the evaluation of the NIOSH research program

Table 1	NORA High-Priority Research Areas by Category
Table 2	Examples of NIOSH Program Research and Transfer Activities
Table 3	Examples of a Variety of Scientific Information Outputs
Table 4	Evaluation Committee Worksheet to Assess Research Programs and Subprograms

ACRONYMS

ABLES	Adult Blood Lead Epidemiology and Surveillance
ACOEM	American College of Occupational and Environmental Medicine
AOEC	Association of Occupational and Environmental Clinics
BLS	Bureau of labor Statistics
CDC	Centers for Disease Control and Prevention
EC	Evaluation Committee
FACE	Fatality Assessment Control and Evaluation
FC	Framework Committee
HHE	Health Hazard Evaluations
MSHA	Mine Safety and Health Administration
NEISS	National Electronic Injury Surveillance System
NIOSH	National Institute for Occupational Safety and Health
NORA	National Occupational Research Agenda
NORA1	National Occupational Research Agenda 1996-2005
NORA2	National Occupational Research Agenda 2005-forward
OSHA	Occupational Safety and Health Administration
OSHAct	Occupational Safety and Health Act of 1970
OSH Review Commission	Occupational Safety and Health Review Commission
PART	Performance Assessment Rating Tool
PEL	Permissible Exposure Limits
SENSOR	Sentinel Event Notification System of Occupational Risks
TMT	Tools, Methods, or Technologies

In September 2004, the National Institute for Occupational Safety and Health (NIOSH) contracted with the National Academies to conduct a review of NIOSH research programs. The goal of this multiphase effort is to assist NIOSH in increasing the impact of its research efforts in reducing workplace illnesses and injuries and improving occupational safety and health. The National Academies agreed to conduct this review and assigned the task to the Division on Earth and Life Studies and the Institute of Medicine.

The National Academies appointed a committee of 14 members, including persons with expertise in occupational medicine and health, industrial health and safety, industrial hygiene, epidemiology, civil and mining engineering, sociology, program evaluation, communication, and toxicology; representatives of industry and of the workforce; and a scientist experienced in international occupational-health issues. The Committee on the Review of NIOSH Research Programs, referred to as the Framework Committee (FC), held meetings during 2005 on May 5-6 and July 7-8 in Washington, DC, and on August 15-16 in Woods Hole and Falmouth, MA.

This document is not a report of the National Academies; rather, it presents the evaluation framework developed by the FC to serve as a guideline and structure for NIOSH program reviews by Evaluation Committees (ECs) to be appointed by various divisions and boards of the National Academies. The ECs will use this framework in reviewing as many as 15 NIOSH research programs during a 5-year period. This is a working document. It is shared with NIOSH and the public. The framework and criteria may be modified by the FC on the basis of responses it receives from the ECs and other sources. It is incumbent upon the ECs to consult with the FC if portions of the evaluation framework presented here are inappropriate for the specific program under review.

I. OVERVIEW OF CHARGE

At the first meeting of the FC, Lewis Wade, NIOSH senior science advisor, emphasized that the reviews should focus on evaluating NIOSH's research programs impact and relevance to health and safety in the workplace. In developing a framework, the FC was asked to address the following:

1. Evaluation committee assessment of progress in reducing workplace illnesses and injuries facilitated by occupational safety and health research through (a) an analysis of relevant data about workplace illnesses and injuries for the program activity, and (b) an evaluation of the effect that NIOSH research has had in reducing illnesses and injuries. The evaluation committees will rate the performance of each

program for impact of the program in the workplace. Impact may be assessed directly or, as necessary, using intermediate outcomes to estimate impact. Qualitative narrative evaluations may also be appropriate under certain circumstances.

2. Evaluation committee assessment of progress in targeting new research to the areas of occupational safety and health most relevant to future improvements in workplace protection.
3. Evaluation committee identification of significant emerging research areas which appear especially important in terms of their relevance to the mission of NIOSH.

Those three charges constitute the scope of work of the individually appointed, independent ECs formed by the National Academies.

I.A. NIOSH Strategic Goals and Operational Plan

As a prelude to understanding the NIOSH strategic goals and operational plan, NIOSH research efforts should be understood in the context of the Occupational Safety and Health Act (OSHAct) under which it was created. The OSHAct identifies workplace safety and health to be a national priority and gives employers the responsibility for controlling hazards and preventing workplace injury and illness. The act creates an organizational framework for doing this, with complementary roles and responsibilities assigned to employers and employees, OSHA, the States, the OSH Review Commission, and NIOSH. As one component of a national strategy the act recognizes NIOSH's roles and responsibilities to be supportive and indirect—NIOSH's research, training programs, criteria and recommendations are all intended to be used to inform and assist those actually responsible for hazard control (OSHAct Section 2b and Sections 20 and 22).

Section 2b of the OSHAct describes thirteen interdependent means of accomplishing the national goal, one of which is "by providing for research . . . and by developing innovative methods . . . for dealing with occupational safety and health problems." Sections 20 and 22 give the responsibility for this research to NIOSH. In addition, NIOSH is given related responsibilities including: the development of criteria to guide prevention of work-related injury or illness, development of regulations reporting on the employee exposures to harmful agents, the establishment of medical examinations programs or tests to determine illness incidence and susceptibility, publication of a list of all known toxic substances, the assessment of potentially toxic effects or risk associated with workplace exposures in specific settings, the conduct of education programs for relevant professionals to carry out the OSHAct purposes, and assisting the Secretary of Labor regarding education programs for employees and employers in hazard recognition and control.

The NIOSH mission is "to provide national and world leadership to prevent work-related illness, injury, disability, and death by gathering information, conducting scientific research, and translating the knowledge gained into products and services". To fulfill its mission, NIOSH has established the following strategic goals:[2]

- **Goal 1: Conduct research to reduce work-related illnesses and injuries.**
  - ♦ Track work-related hazards, exposures, illnesses, and injuries for prevention.
  - ♦ Generate new knowledge through intramural and extramural research programs.
  - ♦ Develop innovative solutions for difficult-to-solve problems in high-risk industrial sectors.
- **Goal 2: Promote safe and healthy workplaces through interventions, recommendations, and capacity-building.**
  - ♦ Enhance the relevance and utility of recommendations and guidance.
  - ♦ Transfer research findings, technologies, and information into practice.
  - ♦ Build capacity to address traditional and emerging hazards.
- **Goal 3: Enhance global workplace safety and health through international collaborations.**
  - ♦ Take a leadership role in developing a global network of occupational health centers.
  - ♦ Investigate alternative approaches to workplace illness and injury reduction and provide technical assistance to put solutions in place.
  - ♦ Build global professional capacity to address workplace hazards through training, information sharing, and research experience.

In 1994, NIOSH embarked on a national partnership effort to identify research priorities to guide occupational health and safety research for the next decade. The National Occupational Research Agenda (NORA) identified 21 high-priority research areas (see Table 1). NORA was intended not only for NIOSH but for the entire occupational health community. Approaching the 10-year anniversary of

[2]See also *http://www.cdc.gov/niosh/docs/strategic/*.

TABLE 1 NORA High-Priority Research Areas by Category

Category	Priority Research Area
Disease and injury	Allergic and irritant dermatitis
	Asthma and chronic obstructive pulmonary disease
	Fertility and pregnancy abnormalities
	Hearing loss
	Infectious diseases
	Low-back disorders
	Musculoskeletal disorders of upper extremities
	Trauma
Work environment and workforce	Emerging technologies
	Indoor environment
	Mixed exposures
	Organization of work
	Special populations at risk
Research tools and approaches	Cancer research methods
	Control technology and personal protective equipment
	Exposure-assessment methods
	Health-services research
	Intervention-effectiveness research
	Risk-assessment methods
	Social and economic consequences of workplace illness and injury
	Surveillance research methods

NORA, NIOSH is working with its partners to update the research agenda. In the second decade of NORA, an approach based on industry sectors will be pursued. NIOSH and its partners will form sector research councils that will work to establish sector-specific research goals and objectives. Emphasis will be placed on moving research to practice in workplaces through sector-based partnerships.

Figure 1 is the NIOSH operational plan presented as a logic model[3] of the path from inputs to outcomes for each NIOSH research program. The FC adapted the model to develop its framework. NIOSH will provide similar logic models relevant to each research program evaluated by an EC.

I.B. Information from Other Evaluations

The FC is aware that several NIOSH programs have already been subjected to evaluation by internal and external bodies. Those evaluations range from overall assessments of NIOSH, such as the recent 2005 Performance Assessment Rating

[3]Developed by NIOSH with the assistance of the RAND Corporation.

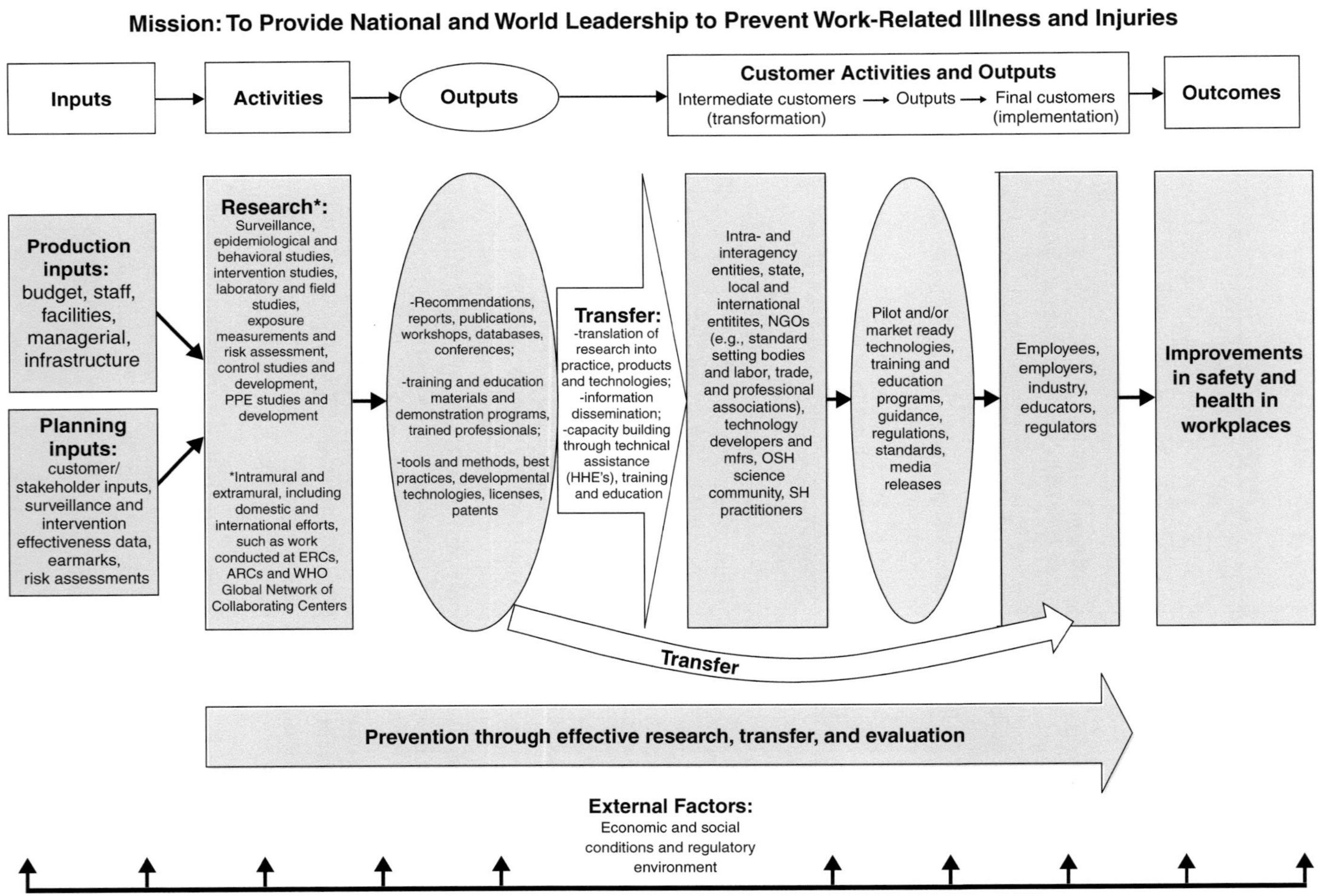

FIGURE 1 The NIOSH operational plan presented as a logic model

Tool (PART) review,[4] to evaluation of research program elements such as any external scientific program reviews. The ECs should review all available prior reviews. Although it is important to consider all prior reviews in the present evaluation to aid in understanding the evolution of the programs and program elements, the ECs' evaluations of NIOSH's programs are independent of the prior reviews and evaluations.

I.C. Evaluation Committees

Individual ECs will be formed through a process consistent with the rules of the National Academies for the formation of balanced committees. The committees will be composed of persons with expertise appropriate to evaluating specific NIOSH research programs and may include representatives of stakeholder groups (such as labor unions and industry) and experts in technology transfer and program evaluation. The committees will conduct appropriate information-gathering sessions to obtain information from the sponsor (a NIOSH research program), stakeholders affected directly by the NIOSH research, and relevant independent parties. Each EC will consist of about 10 members, will meet about three times, and will prepare a report. The National Academies will deliver the report to NIOSH within 9 months after the individual EC is formed. EC reports will be subjected to the National Academies report-review process.

I.D. Evaluation Committees' Information Needs

The ECs are expected to conduct information-gathering as appropriate on

- Background and resources of the program:
    - History of program, including results of previous reviews.
    - Program funding, by year, for the current year and the last 10 years.
    - Program funding, by objective or subprogram.
    - Extramural-grant awarding, cooperative agreement and contracting process, solicitation of research ideas, and advisory activities.
- Program goals and objectives.
- Internal NIOSH processes and research:
    - Intramural surveillance, research, and transfer activities.
    - Process to solicit and approve intramural research proposals.

[4]PART focuses on assessing program-level performance and is one of the measures of success for the Budget and Performance Integration initiative of the president's management agenda (see CDC Occupational Safety and Health at *http://www.whitehouse.gov/omb/budget/fy2006/pma/hhs.pdf*).

- NIOSH-funded extramural research:
    - Requests for proposals, cooperative agreements and research contracts distributed.
    - Awardee products, including close-out reports, surveillance, research, and transfer activities, peer-reviewed publications, and patents.
- Products and technology transfer:
    - Data related to program publications, conferences, recommendations, patents, and so on.
    - Past and planned mechanisms for transferring outputs to outcomes.
    - Interventions, recommendations, and information-dissemination and technology-transfer activities designed to get research findings used to improve occupational safety and health.
    - Outcomes of research, alerts, standard-setting, investigations, and consultations; for example—documented reductions in risk after program-supported interventions, employer and industry behavior changes made in response to research outputs, and worker behavior changes in response to research outputs.
- Impact on worker safety and health—data necessary to evaluate program impact on health outcomes (work-related injuries and illnesses) and exposures.
- The most severe or most frequent adverse health and safety outcomes or exposures in the research program and the most accessible improvements with respect to health and safety.
- Interactions within NIOSH and with other stakeholders:
    - The role of program research staff in NIOSH policy-setting, Occupational Safety and Health Administration (OSHA) and Mine Safety and Health Administration (MSHA) standard-setting, and voluntary standard-setting and other government policy functions.
    - Other institutions and research programs with overlapping or similar portfolios and an explanation of the relationship between the NIOSH work and staff and those of other institutions.
    - Stakeholder perspectives (OSHA, MSHA, union and workforce, industry, and so on.)
    - Key partnerships with employers, labor, other government organizations, academic institutions, nonprofit organizations, and
    - International involvement and perspective.
- Systems to identify emerging problems and emerging research, including plans.

II. SUMMARY OF EVALUATION PROCESS

The ECs are charged with assessing the relevance, quality, and impact of NIOSH research programs. In conducting their evaluations, the ECs should ascertain whether NIOSH is doing the right things (relevance) and doing them right (quality) and whether these things are improving health and safety in the workplace (impact).

II.A. The Evaluation Flow Chart (Figure 2)

To address its charges, the FC has developed a flow chart (Figure 2) that breaks the NIOSH logic model into discrete, sequential program components to be characterized or assessed by the ECs. The components to be assessed are as follows:

- Major program-area *challenges.*
- Strategic *goals and objectives.*
- *Inputs* (such as budget, staff, facilities, the institute's research management, the NIOSH Board of Scientific Counselors, the NORA process, and NORA work groups).
- *Activities* (efforts by NIOSH staff, contractors, and grantees, such as hazard and health-outcome surveillance, exposure-measurement research, health-effects research, intervention research, health services, other research, and technology-transfer activities).
- *Outputs* (the products of NIOSH activities, such as publications, reports, conferences, databases, tools, methods, guidelines, recommendations, education and training, and patents).
- *Intermediate outcomes* (responses by NIOSH stakeholders to NIOSH products, such as public or private policy change, training and education in the form of workshop or seminar attendance, self-reported use or repackaging of NIOSH data by intermediary stakeholders, adoption of technologies developed by NIOSH, implemented guidelines, licenses, and reduction of workplace hazardous exposures and other risk factors).
- *End outcomes* (such as reduction of work-related injuries or illnesses, or hazardous exposures in the workplace).

Drawing on the program logic model, the flow chart, and EC members' expertise, the ECs will delineate important determinants of a NIOSH research program's agenda and the consequences of the NIOSH research activity. Determinants are conceptualized as inputs and external factors. Examples of external factors are

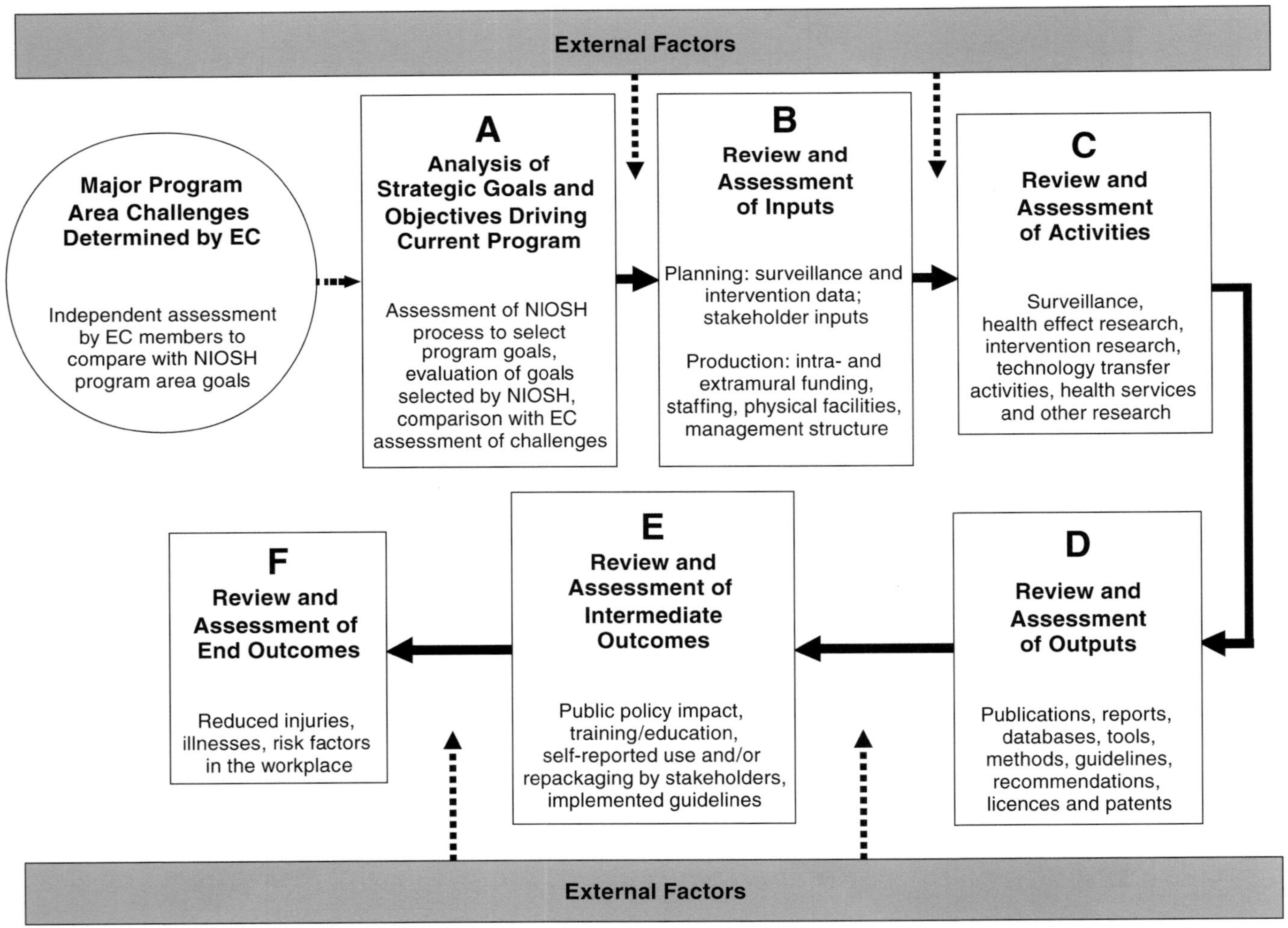

FIGURE 2 Flow chart for the evaluation of the NIOSH research program

the research activities of industry and other federal agencies and the political and regulatory environment, which can affect all components of the research program (Figure 2). *For purposes of this review, the results of inputs and external factors are the program research activities, outputs, and associated transfer activities that may result in intermediate outcomes and possibly eventual end outcomes.*

The FC has used the NIOSH logic model to develop the flow chart to define the scope and steps of an EC evaluation. The FC's vision of how a program evaluation should occur is incorporated in a summary manner in the flow chart and discussed extensively in later sections. For example, the FC identified two types of outcomes: (a) *intermediate outcomes*, which represent implementations (what external stakeholders, such as employers, do in reaction to the products of NIOSH work, including new regulations, widely accepted guidelines, introduction of control technologies in the workplace, changes in employer or worker behaviors, and changes in diagnostic practices of health-care providers), and (b) *end outcomes*, which are improvements (reductions in work-related injuries, illnesses, and hazardous exposures). For the purpose of evaluation, the FC does not differentiate between NIOSH's "intermediate customer" and "final customer" activities (Figure 1); instead it combines them into a single category (Box E, Review and Assessment of Intermediate Outcomes, Figure 2). Training and development programs were appropriately defined as outputs by NIOSH in the logic model, but the FC finds more value in focusing on response to such offerings as intermediate outcomes (Box E) in the flow chart. The number of workers exposed to training activities represents a type of implementation of NIOSH outputs in the workplace. In evaluating each program or major subprogram, the EC must collect, analyze, and evaluate information on items described in each of the boxes of Figure 2. Further details on the evaluation are described in Section III of this document.

II.B. Steps in Program Evaluation

The FC has concluded that useful evaluation requires: (a) a disciplined focus on a small number of questions or hypotheses typically related to program goals, performance criteria, and performance standards; (b) a rigorous method for answering the questions or testing the hypotheses; and (c) a credible procedure for developing qualitative and quantitative assessments. The evaluation process developed by the FC is summarized here and described in detail in Section III of this document.

1. Gather appropriate information from NIOSH and other sources.
2. Determine timeframe that the evaluation will cover (see III.B.1).
3. Identify program-area major challenges and objectives (see III.B.2). All NIOSH research programs, whether based on health outcomes or sectors, are designed to be responsive to the safety and health problems

in today's or tomorrow's workplace. In the NIOSH vision, mission, values, and goals, each research program should have its own objectives. The ECs will provide an independent assessment of the major program challenges and determine whether they are consistent with the research program's stated goals and objectives.

4. Identify subprograms and major projects in the research program. It is important for each EC to determine how necessary it is to disaggregate a program to achieve a manageable and meaningful evaluation of its components and the total program. Each research program may need to be broken down into several recognizable subprograms or major projects if an effective evaluation is to be organized. It may be advantageous for an EC to disaggregate a program into subprograms that NIOSH identifies.
5. Evaluate the program and subprogram components sequentially as discussed in Section III, using the flow chart (Figure 2) as a guide (Sections III.B.3 through III.B.8). This will involve qualitatively assessing each phase of a research program by using the questions and guidance provided by the FC and professional judgment.
6. Evaluate the research program's potential outcomes not yet appreciated (Section III.B.9).
7. Evaluate and score the program outcomes and important subprogram outcomes specifically for contributions to improvements in workplace safety and health. A worksheet is provided with specific items for consideration (Section III.B.10).
8. Evaluate and score the overall program for impact (Section III.B.10). Final program ratings will consist of a numerical score and discussion of its rationale.
9. Evaluate and score the overall program for relevance (Section III.B.10). Final program ratings will consist of a numerical score and discussion of its rationale.
10. Identify significant emerging research areas (Section III.C). On the basis of the expert judgment of EC members and information gathered from stakeholders (such as, labor, industry, academe, and government agencies) and from appropriate NIOSH sentinel-event field-investigation activities, the EC will respond to Charge 3 by identifying and describing emerging research that appears especially important in its relevance to the mission of NIOSH. The EC will assess the extent to which NIOSH's program is responsive to today's and tomorrow's needs and determine whether there are any gaps in response.
11. Prepare report by using the template provided in Section IV as a guide.

II.C. Assessing Relevance

FC members identified numerous *possible* factors to consider in assessing the relevance of NIOSH research programs, such as:

- The severity, frequency, or both of the health and safety outcomes addressed and the number of people at risk (magnitude) for these outcomes.
- The extent to which NIOSH research programs have identified and addressed gender issues and the concerns related to vulnerable populations. Vulnerable populations are defined as groups of workers who have (1) biological, social, or economic characteristics that place them at increased risk of developing work-related conditions and/or (2) inadequate data collected about them. Vulnerable populations include disadvantaged minorities, disabled individuals, low-wage workers, and non-English speakers for whom language or other barriers present health or safety risks.
- The extent to which NIOSH research programs have addressed the health and safety needs of small businesses.
- The "life stage" of the problems being addressed. As the health effects are understood, emphasis should shift to intervention research, and from efficacy to effectiveness to research on the process of dissemination of tested interventions. Gaps in the spectrum of prevention need to be addressed; for example, research on exposure assessment may be necessary before the next intervention steps can be taken.
- The structure, in addition to the content, of the research program. A relevant research program is more than a set of unrelated research projects; it is an integrated program involving an interrelated set of surveillance, research, and transfer activities.
- Appropriate consideration by NIOSH of stakeholder inputs.

II.D. Assessing Impact

Causal attribution is a major aspect of program evaluation. It is necessary for the ECs to assess, to the extent possible, NIOSH's contribution to end outcomes. Data on reductions in work-related injuries, illnesses, and hazardous exposures will be available for some programs. In some cases, they may be quantifiable. It is possible, however, to evaluate the impact of a NIOSH research program whether the outcomes are intermediate outcomes or end outcomes. Intermediate outcomes may be used as proxies for end outcomes in assessing impact if there is no direct evidence of improvements in health and safety as long as the ECs qualify their findings. The ECs will describe the realized or potential benefits of NIOSH's programs. Examples of realized intermediate outcomes include: new regulations, widely accepted guidelines, work practices, and procedures, all of which may contribute measurably to enhancing health and safety at the work place.

The contribution of a NIOSH program to technology now in use or being implemented is another important part of impact assessment. NIOSH's contribution can be assessed as major or important, moderate, likely, limited, or none. If technology development is in progress or has been abandoned, for whatever reason, the benefits are only potential or consist of knowledge gain.

III. EVALUATION OF NIOSH RESEARCH PROGRAMS—THE PROCESS

III.A. Analysis of External Factors Relevant to the NIOSH Research Program

As depicted in the logic model (Figure 1), the end outcome of reduced injuries, illnesses, or exposures is effected through stakeholder activities and outputs. All those involve the use of NIOSH outputs by stakeholders in industry, labor, other government agencies, and so on. It is evident that actions beyond NIOSH's control—by industry, labor, and other entities—have important bearings on the incorporation in the workplace of NIOSH's outputs to enhance health and safety. The implementation of research findings may depend on existing or future policy considerations.

III.A.1. Overview

External factors may be considered as forces beyond the control of NIOSH that may affect the evolution of the program. External factors dominate the evolution of the path from NIOSH inputs to occupational health and safety outcomes (Figure 1). External factors can also be considered inputs to the evaluation of each aspect (planning, implementation, transfer, and others) of NIOSH research programs (Figure 2).

Identification of external factors by the ECs is essential to providing a context for NIOSH program evaluation. External factors may best be assessed through the expert judgment of EC members regarding the knowledge base, the research program, and implementation of interventions as these relate to the needs in the occupational health or safety area targeted by the research program. The ECs, however, may choose additional approaches to assess external factors.

The FC recommends the ECs ask NIOSH to identify and describe external factors early in the evaluation sequence. Factors external to NIOSH might have been responsible for achieving some outcomes, and they might also have presented formidable obstacles. The ECs must address both possibilities.

III.A.2. Considerations for Discussion

Some external factors may involve constraints on research activity related to target populations, methodological issues, and resource availability. For example, evaluators might examine whether

- Projects addressing a critical health need are technologically feasible. A workforce with appropriate size and duration, magnitude, and distribution of exposure for measuring a health effect may not exist. For example, no population of workers has been exposed for 30 years to formaldehyde at the current OSHA Permissible Exposure Level (PEL), so the related cancer mortality can not yet be directly assessed.
- Research is inhibited because NIOSH investigators are unable to access an adequate study population. Under current policy, NIOSH must either obtain an invitation by management to study a workplace or seek a judicial order to provide authority to enter a worksite. (Cooperation under court order may well be insufficient for effective research.)
- Research is inhibited because the work environment, materials, and historical records cannot be accessed even with management and workforce cooperation.
- Adequate or established methods do not exist for assessing the environment.
- Records needed for historical-exposure reconstruction cannot be accessed or do not exist.
- Intervention research is inhibited because an appropriate employer partner cannot be identified to institute the intervention.
- The NIOSH contribution to a certain area of research is reduced because other institutions are working in the same area.
- NIOSH resources are inadequate to tackle the key questions.

Evaluation of the impact of NIOSH research outputs on outcomes may require consideration of external factors that might have impeded or aided implementation, measurement, and so on. For example, evaluators might consider whether

- Regulatory end points are unachievable because of obstacles to regulation or differing priorities of the regulatory agencies. For example, recommendations for improved respiratory protection programs for health-care workers might not be implemented because of enforcement policies or lack of acceptance by the administration of health-care institutions.
- A feasible control for a known risk factor or exposure is not im-

plemented because the costs of implementation are too high or the economic incentives under current circumstances do not favor such actions.

- Improvements in end points are unobservable because baseline and ongoing surveillance data are not available. For example, the current incidence of occupational noise-induced hearing loss is not known although surveillance for a significant threshold shift is feasible. (NIOSH conducts surveillance of work-related illnesses, injuries, and hazards, but comprehensive surveillance is not possible with existing resources.)
- Reductions in adverse effects of chronic exposure cannot be measured. For example, 90% of identified work-related mortality is from diseases, such as cancer, that arise only after decades of latency from first exposure; therefore, effects of reducing exposure to a carcinogen cannot be observed in the timeframe of most interventions.
- A regulation is promulgated that requires a technology that was developed but not widely used.

III.B. Evaluating NIOSH Research Programs (Addressing Charges 1 and 2)

III.B.1. Identifying Period of Time to Be Evaluated

Through study of materials presented by the NIOSH research program and other sources, an EC will become familiar with the history of the research program being evaluated and its major subprograms, program goals and objectives, resources, and other pertinent information.

It is useful for the ECs to consider three general timeframes in conducting their reviews:

- 1970-1995, the period from the founding of NIOSH to the initiation of the NORA process (pre-NORA period).
- 1996-2005 (NORA 1 period).
- Current period and forward (NORA 2 period).

It will be important for the ECs to get a general sense of the history of the NIOSH research program and its impact, but their efforts should be focused on the impact and relevance of NIOSH programs from 1996 on. It is recognized that many of the intermediate and end outcomes since 1996 are the consequence of research outputs accomplished earlier. Both the relevance of the research program targets of NORA 1 and the proposed NORA 2 objectives for the next decade should be considered.

NIOSH is in the midst of a substantial restructuring of the NORA agenda, and expert judgment about relevance and prospective impact of current research programs will be most useful to the agency. The timeframes provided here are only for general guidance; the exact dates of the period to focus on in reviewing programs will depend on the specific research program under review.

III.B.2. Identification of Major Challenges (Circle in Figure 2)

Early in its assessment process, an EC should independently identify the major challenges for its research program. These would be the matters the EC believes should have priority in the research program being evaluated. In arriving at a list of challenges, the EC should rely on surveillance findings, including NIOSH investigations of sentinel events (through health-hazard or fatality-assessment programs), and its own expert judgment. Those should be supplemented with determinations or recommendations by appropriate advisory sources regardless of whether these sources have contributed to NIOSH program deliberations. This process will allow the EC to compare its assessment of challenges to be addressed by NIOSH with NIOSH program goals, and to evaluate the congruence between the two as a measure of relevance (Charge 2).

III.B.3. Analysis of Research Program Strategic Goals and Objectives (Box A in Figure 2)

The research program goals and objectives should be evaluated, with a focus on how each research program's goals are related to NIOSH's agency-wide strategic goals and to the major current challenges and emerging problems identified in the step above. Differences may exist between the importance or relevance of an issue and the influence NIOSH-funded research might have in addressing the issue. The EC should recognize that NIOSH research priorities may be strategic rather than based on the assessment of the state of knowledge.

Some aspects of the NIOSH research program's strategic goals and objectives would have been already subjected to evaluation by internal or external bodies. Research program relevant evaluations that should be requested include the NIOSH annual program review by the Leadership Team; the NORA research program proposal pre-award external review, NORA post-award program external review, and external scientific program review.

Questions to Guide the Evaluation Committee

1. Are the strategic goals and objectives of the program well defined and clearly described?
2. In the last decade, how well were program goals and objectives aligned with NORA 1 priorities?
3. How do the current strategic goals and objectives of the program relate to the current NIOSH strategy, including NORA 2?
4. Are the research program goals, objectives, and strategies relevant to the major challenges in the research program and likely to address emerging problems in the research program (as determined by the EC)?
   a. Did past program goals and objectives (research and dissemination/transfer activities) focus on the most relevant problems and anticipate the emerging problems in the research program?
   b. Are the current program goals and objectives targeted to the most relevant problems and likely to address emerging problems in the research program?
5. How does the program identify emerging research areas?
   a. What information is reviewed by NIOSH?
   b. What advisory or stakeholder groups are asked to identify emerging areas?
   c. What new research areas have been identified in the program?
   d. Were important areas overlooked?

Assessment

The EC will provide a qualitative assessment discussing the relevance of the area's goals, objectives, and strategies as related to the research program's major challenges and emerging problems.

III.B.4. Review of Inputs (Box B in Figure 2)

Inputs are categorized as planning or production inputs in the NIOSH logic model. Planning inputs include stakeholder inputs, surveillance and intervention data, and risk assessments. Production inputs include intramural and extramural funding, staffing, management structure, and physical facilities.

Inputs for program evaluation include existing intramural and extramural information and, potentially, surveys or case studies that might have been developed specifically to assess progress in reducing workplace illnesses and injuries and to provide information relevant to targeting research appropriately to future needs. The ECs should request the relevant planning and production inputs from NIOSH.

Planning Inputs

Planning inputs can be qualitative or quantitative. Sources of qualitative inputs include

- Federal Advisory Committee Act panels (Board of Scientific Counselors, Mine Safety and Health Research Advisory Committee, National Advisory Committee on Occupational Safety and Health, and so on).
- NORA research partners, initial NORA stakeholder meetings, later NORA Team efforts (especially strategic research plans), and the NORA Liaison Committee and federal liaison committee recommendations.
- Other federal research agendas, industry, labor, academe, professional associations, industry associations, and Council of State and Territorial Epidemiologists.
- OSHA and MSHA strategic plans.

Attention should be given to how comprehensive the inputs have been and to what extent gaps have been identified or considered.

Sources of quantitative inputs include

- Intramural surveillance information, such as descriptive data on exposures and outcomes (appropriate data may be available from a number of NIOSH divisions and laboratories).
- Health Hazard Evaluations (HHEs).
- Reports from the Fatality Assessment Control and Evaluation (FACE) program.
- Extramural health-outcome and exposure-assessment data from (1) OSHA and MSHA (inspection data) and the Bureau of Labor Statistics, U.S. Department of Defense, and U.S. Department of Agriculture (fatality, injury, and illness surveillance data); (2) state government partners, including NIOSH-funded state surveillance programs, such as Sentinel Event Notification System of Occupational Risks (SENSOR), Adult Blood Lead Epidemiology and Surveillance (ABLES), and state-based FACE; and (3) non-government organizations, such as the Association of Occupational and Environmental Clinics (AOEC) and

the American College of Occupational and Environmental Medicine (ACOEM).

- Appropriate data from NIOSH-funded, investigator-initiated extramural research.

Production Inputs

For each research program under review, NIOSH should specify an identifiable portion of the NIOSH intramural budget, staff, facilities, and management that has been allocated by divisions and offices that play a major role in the research program. Production inputs should be described primarily in terms of intramural research projects and staff, relevant extramural projects (particularly cooperative agreements and contracts), and HHEs and related staff. Consideration should also be given to budget inputs for program evaluation and to leveraged funds provided by partners, such as National Institutes of Health and the Environmental Protection Agency joint requests for applications or program announcements and OSHA, MSHA, and Department of Defense contracts with NIOSH to conduct work.

Assessment of those inputs should include consideration of (1) the degree to which allocation of funding and personnel has been reasonably consistent with the resources needed to conduct the research and (2) the extent to which funding for the relevant intramural research program activity has been limited by lack of discretionary spending beyond salaries (travel, supplies, external laboratory services, and so on). The assessments, therefore, should consider the adequacy of the qualitative and quantitative planning inputs and the use and adequacy of production inputs, particularly (1) and (2) above.

Questions as a Guide for the Evaluation Committee

1. Were the planning, production, and other input data adequate?
2. How well were the major planning, production, and other program inputs used to promote the major activities?
3. Were the sources of inputs and the amount and quality of inputs adequate?
4. Was input obtained from stakeholders representing vulnerable working populations and small businesses?
5. Were production inputs (intramural and extramural funding, staffing, management, and physical infrastructure resources) consistent with goals and objectives of the program?

Assessment

The EC will provide a qualitative assessment that discusses the quality, adequacy, and use of inputs.

III.B.5. Review of Activities (Box C in Figure 2)

Activities are defined as the efforts and work of the program, its staff, and its grantees and contractors. For purposes of the present evaluation, activities of the NIOSH program under review should be divided into research and transfer activities. Research activities may be further categorized as surveillance, health-effects research, intervention research, health-services research, and other research (see sample classification of research activities in Table 2). Transfer activities include information dissemination, training, technical assistance, and education designed to translate research outputs into content and formats designed for application in the workplace to produce improvements in occupational safety and health. Depending on the scope of the program under review, activities may also be grouped by research program objectives or subprograms.

Conventional occupational-health research focuses appropriately on health effects and technology. A focus on socioeconomic and policy research and on surveillance and diffusion research is also needed to effect change because not all relevant intermediate outcomes occur in the workplace. There are important outcomes farther out on the causal chain that NIOSH can affect and thereby influence health and safety in the workplace. Some examples of types of research that might also prove important in addressing NIOSH's mission are

- Socioeconomic research on cost shifting between worker compensation and private insurance.
- Surveillance research to assess the degree of significant and systematic underreporting of select injuries and illnesses on OSHA logs.
- Research on methods to build health and safety capacity in community health centers that serve low-income and/or minority-group workers, and to improve recognition and treatment of work-related conditions.
- Transfer research to change health and safety knowledge in teenagers while they are in high school to improve the likelihood of reduced injuries when they enter the workforce.
- Community-based participatory research on differences between recently arrived immigrants and US-born workers regarding perceptions of acceptable health and safety risks to target programs to meet the workforce training needs of immigrant workers.

TABLE 2 Examples of NIOSH Program Research and Transfer Activities

Surveillance
(including hazard and health surveillance and evaluation of surveillance systems)

Health-effects research
Epidemiologic research
Toxicologic research
Laboratory-based physical and safety risk factor research
Development of clinical screening methods and tools

Exposure-assessment research
Intervention research
Control technology
  Engineering controls and alternatives
  Administrative controls
  Personal protective equipment
Work organization research
Community-based participatory research
Policy research (such as alternative approaches to targeting inspections)
Diffusion and dissemination research
  Training effectiveness
  Information-dissemination effectiveness
  Diffusion of technology

Health-services and other research
Access to occupational health care
Infrastructure research—delivery of occupational-health services, including international health and safety
Socioeconomic consequences of work-related injuries and illnesses
Worker compensation

Technology-transfer and other transfer activities
Information dissemination
Training programs

The ECs should review the list of research and transfer activities (projects) for the research program under review that have been completed, are in progress, or have been planned. Surveillance activities should be included in this review. An EC should request that the NIOSH program under review provide a list of activities, grouping the projects into research activities as in Table 2, and specify whether they are intramural or extramural. For extramural projects, the key organizations and principal investigators' names should be requested, as should whether the projects were in response to a request for proposal or a request for application. For an intramural project, the EC should ask NIOSH to provide a list of key collaborators (other government agency, academe, industry, and/or union partners).

The ECs should evaluate each of the research activities outlined in Table 2 to the extent that each forms an important element of the program research. In the case of a sector research program (for example, mining, construction) in which health-effects research is not being reviewed, the ECs should determine what research inputs are being used by the program to develop its targets and then assess the value of the inputs.

Questions to Guide the Evaluation Committee in Assessing Research Activities

1. What are the major subprograms or groupings of activities within the program?
2. Were the activities consistent with program goals and objectives?
3. Were the research activities relevant to the major challenges in the research program?
   a. Did they address the most serious outcomes?
   b. Did they address the most common outcomes?
   c. Did they address the needs of both genders, vulnerable working populations, and small businesses?
4. Were the research activities appropriately responsive to the input of stakeholders?
5. To what extent were partners involved in the research activities?
6. Are the resource allocations appropriate, and appropriate at this time, for the research activities?
7. To what extent did peer reviews (internal, external, and precourse or midcourse) affect the activities?
8. Is there adequate monitoring of quality assurance procedures to ensure credible research data, analyses, and conclusions?

Questions to Guide the Evaluation Committee in Assessing Transfer Activities

1. Is there a coherent planned program of transfer activities?
2. Are the program's information dissemination, training, education, technical assistance, or publications successful in reaching the workplace or relevant stakeholders in other settings? How widespread is the response?
3. To what extent did the program build research and education capacity (internal or external)?

Assessment

For this part of the assessment, the EC will provide a qualitative assessment discussing relevance and quality. This evaluation must include consideration of the external factors identified in Section III.A that constrain choices of research projects. The EC will consider the appropriateness of resource allocations with respect to issues' importance and the extent to which the issue is being addressed. A highly relevant and high-quality program would be comprehensive, address high-priority needs, produce high-quality results, be highly collaborative, and be of value to stakeholders. Programs may be progressively less relevant or of lower quality as those key elements are not up to the mark or are missing. The discussion should cover those aspects in sufficient detail to arrive at a qualitative assessment of the activities. Assessment of the transfer activities must include considerations of program planning, coherence, quality, and impact.

III.B.6. Review of Outputs (Box D in Figure 2)

As shown in Figure 1, research inputs and activities lead to outputs. An output is a direct product of a NIOSH research program that is logically related to the achievement of desirable and intended outcomes. Outputs are created for researchers, practitioners, intermediaries, and end-users, such as consumers. Outputs can be in the form of publications in peer-reviewed journals, recommendations, reports, Web-site content, workshops and presentations, databases, educational materials, scales and methods, new technologies, patents, technical assistance, and so on. Outputs of NIOSH's extramurally funded activities should also be considered. Examples of major outputs are provided in Table 3.

Depending on the intended audience, outputs may be tailored to communicate information most effectively to increase the likelihood of comprehension, knowledge, attitude formation, and behavioral intent. The extent of use of formative evaluation data (data gathered prior to communication for the purpose of improving the likelihood of the intended effects) or intended user feedback in the design of the output can be considered an indicator of output quality.

In addition to outputs themselves, many related indicators of the production, reference to, and utility of outputs can be conceptualized and made operational. Examples include the extent of collaboration with other organizations in the determination of research agendas, the conduct of research, the dissemination of research results, and interorganizational involvement in the production of outputs. Coauthorship is a measure of the centrality of NIOSH researchers in the broader research community.

TABLE 3 Examples of a Variety of Scientific Information Outputs

Peer-reviewed publications by NIOSH staff
Total number of original research articles by NIOSH staff
Total number of review articles by NIOSH staff (including best-practice articles)
Complete citation for each written publication
Complete copies of the "top five" articles
Collaboration with other public- or private-sector researchers
Publications in the field of interest with other support by investigators also funded by NIOSH (for example, ergonomic studies with other support by an investigator funded by NIOSH to do ergonomics work, in which case NIOSH should get some credit for seeding interest or drawing people into the field)

Peer-reviewed publications by external researchers funded by NIOSH
Total number of NIOSH-funded original research articles by external researchers
Total number of NIOSH-funded review articles by external researchers (including best-practices articles)
Complete citation for each written report
Complete copies of the "top five" articles
Collaboration with other government or academic researchers

NIOSH reports in the research program
Total number of written reports
Complete citation for each written report
Complete copies of the "top five" reports

Sponsored conferences and workshops
Total number of sponsored conferences
Total number of sponsored workshops
For each sponsored conference or workshop, describe:
- Title, date, and location
- Partial vs complete sponsorship (if partial, who were cosponsors?)
- Approximate number of attendees and composition of participants
- Primary "products" of the event (such as publication of conference proceedings)

NIOSH's assessment of value or impact

Databases
Total number of major databases created by NIOSH staff
Total number of major databases created by external researchers funded by NIOSH grants,
For each database:
- Title, objective (in one to four sentences), and start and stop dates
- Partial vs complete sponsorship (if partial, who were cosponsors?)
- Study or surveillance-system design, study population, and sample size
- Primary "products" of the database (such as number of peer-reviewed articles and reports)
- Complete copies of the "top two" publications and/or findings, to date, from each database

continued

TABLE 3 Continued

Recommendations
Total number of major recommendations
For each:
Complete citation (article, report, or conference where recommendation was made)
Summary in one to four sentences
Percentage of target audience that has adopted recommendation 1, 5, and 10 years later
Up to three examples of implementation in the field
Identifications of "top five" recommendations to date
Tools, methods, or technologies (TMT)
Total number of major TMT (includes training and education materials)
For each:
Title and objective of TMT (in one to four sentences)
Complete citation (if applicable)
Percentage of target audience that has used TMT 1, 5, and 10 years later
Up to three examples of implementation in the field
Identification of "top five" TMT to date
Patents
Total number of patents
For each:
Title and objective patent (in one to four sentences)
Complete citation
Percentage of target audience that has used product 1, 5, and 10 years later
Up to three examples of implementation in the field
Identification of "top five" patents to date
Miscellaneous
Any other important program outputs

The EC should ask NIOSH to provide information on all relevant outputs for the specific program for the chosen time period.

Questions to Guide the Evaluation Committee

1. What are the major outputs of the research program?
2. Did the research program produce outputs that addressed the high-priority areas?
3. To what extent did the program generate important new knowledge or technology?
4. Are there peer-reviewed publications that are widely cited and considered to report "breakthrough" results?

5. Were outputs relevant to both genders, vulnerable populations and health disparities?
6. Were outputs relevant to health and safety problems of small businesses?
7. Are products user-friendly in terms of readability, simplicity, and design?
8. To what extent did the program help to build the internal or extramural institutional knowledge base?
9. Did the research produce effective cross-agency, cross-institute, or internal-external collaborations?

Assessment

For this part of the assessment, the EC should provide a qualitative assessment discussing relevance, quality, and usefulness. A highly ranked program will be one with outputs that address needs in high-priority areas, contain new knowledge or technology that is effectively communicated, contribute to capacity-building both inside and outside NIOSH, and are relevant to the pertinent populations. The discussion should cover those aspects in sufficient detail to support the qualitative assessment of the outputs.

III.B.7. Review of Intermediate Outcomes (Box E in Figure 2)

Intermediate outcomes, for the purposes of this evaluation, are related to the program's association with behaviors and changes at individual, group, and organizational levels in the workplace. An intermediate outcome reflects an assessment of worth by stakeholders outside NIOSH (such as managers in industrial firms) about NIOSH research or its products.

Intermediate outcomes include the production of standards, or regulations based in whole or in part on NIOSH research (products adopted as public policy or as policy or guidelines by private organizations or industry); attendance at training and education programs sponsored by other organizations; use of publications by workers, industry, and occupational safety and health professionals in the field; and citations of NIOSH research by industrial and academic scientists.

More difficult-to-collect intermediate outcomes that may be valid indicators of quality or utility include self-report measures by users and relevant nonusers of NIOSH outputs. These indicators include the extent to which key intermediaries find value in NIOSH databases for the repackaging of health and safety information, the extent to which NIOSH recommendations are in place and attended to in workplaces, and employee or employer knowledge of and adherence to NIOSH recommended practices.

A research program might be evaluated in terms of whether it is recognized as a national center of excellence, is one of the larger and best research programs in the country, is recognized only in terms of particular staff or a particular laboratory, duplicates other, larger facilities, or is not unique or has little capability or capacity.

Questions to Guide the Evaluation Committee

1. Has the program resulted in stakeholder training or education activities that are being used in the workplace or in school or apprentice programs? If so, what is the response to what is being done, and how widespread is the response?
2. Has the program resulted in standards, regulations, public policy, or voluntary guidelines that have been transferred to or created by the workplace in response to NIOSH outputs?
3. Has the program resulted in new control technology or administrative control concepts that are feasible for use or have been adopted in the workplace to reduce risk factors?
4. Has the program resulted in new personal protective equipment that is feasible for use or has been adopted in the workplace to reduce risk factors or exposures?
5. Has the program contributed to changes in health care practices to improve recognition and management of occupational health conditions?
6. Has the program resulted in research partnerships with stakeholders leading to changes in the workplace?
7. To what extent did the program's stakeholders find value in NIOSH's products (as shown by document requests, web hits, conference attendance, and so on)?
8. Has the program resulted in changes in employer or worker practices associated with the reduction of risk factors?
9. Does the program or a subprogram provide unique staff or laboratory capability that is a necessary national resource? If so, is it adequate or does it need to be enhanced or reduced?
10. Has the program resulted in interventions that protect both genders, vulnerable workers or address the needs of small businesses?
11. To what extent did the program contribute to increased capacity at worksites to identify or respond to threats to safety and health?

Assessment

Only a qualitative assessment of product development, usefulness, and impact is required at this point in the EC report. Some thought should be given to the relative value of intermediate outcomes, and the FC recommends applying the well-accepted hierarchy-of-controls model. The discussion could include comments on how widely products have been used or programs implemented. The qualitative discussion should be specific as to the various products developed by the program and the extent of their use by specific entities (industry, labor, government, and so on) for specific purposes. Whether the products have resulted in changes in the workplace or in the reduction of risk factors should be discussed. The recognition accorded to the program or the facilities by its peers (such as recognition as a "center of excellence" by national and international communities) should be considered in the assessment. A program to be highly ranked should have a high level of performance in most of the relevant questions in this section. Whether the impact was caused by NIOSH alone or in combination with external agents should also be considered in the evaluation. An aspect of the evaluation can be whether the impact would have probably occurred without NIOSH's efforts.

III.B.8. Review of End Outcomes (Box F in Figure 2)

End outcomes are defined by measures of health and safety and of impact on process and programs. The FC recognizes that a major challenge in assessing the causal relationship between NIOSH research and specific occupational health and safety outcomes is that NIOSH does not have direct responsibility or authority for implementing its research findings in the workplace. Furthermore, the benefits of NIOSH research program outputs can be realized, potential, or limited to knowledge gain. For example, negative studies contribute to the knowledge base and the generation of important new knowledge is a recognized form of outcome, in the absence of measurable impacts.

Outcome impact depends on there being a "receptor" for research results, including regulatory agencies, consensus and professional organizations, and employers. The ECs should consider questions related to the various stages that lead to outputs, such as

1. Did NIOSH research identify a gap in protection or a means of reduction of risk?
2. Did NIOSH convey that information to potential users in a usable form?
3. Was the research applied?
4. Did the results work?

End outcomes, for purposes of this evaluation, are changes related to health, including decreases in injuries, illnesses, deaths, and decreases in exposures or risk factors resulting from the research in the specific program or subprogram. Quantitative data are preferable to qualitative, but qualitative analysis may be necessary.

Sources of quantitative data include

- Bureau of Labor Statistics (BLS) data on fatal occupational injuries (Census of Fatal Occupational Injuries) and nonfatal injuries and illnesses (Annual Survey of Occupational Injury and Illnesses).
- NIOSH intramural surveillance systems, such as the National Electronic Injury Surveillance System (NEISS), the coal worker x-ray surveillance program, and agricultural worker surveys conducted by NIOSH in collaboration with the US Department of Agriculture.
- State-based surveillance systems, such as the NIOSH-funded ABLES, and the SENSOR programs (for asthma, pesticides, silicosis, noise-induced hearing loss, dermatitis, and burns).
- Selected state workers-compensation programs.
- OSHA, which collects exposure data, in the Integrated Management Information System.

The FC is unaware of surveillance mechanisms for many occupationally related chronic illnesses such as cancers arising from long exposure to chemicals and other stressors. For many outcomes, incidence and prevalence are best evaluated by investigator-initiated research.

The strengths and weaknesses of the various sources of outcome data should be recognized by the ECs. Quantitative accident, injury, illness, and employment data and databases are subject to error and bias and should be used by the ECs for drawing inferences only after critical evaluation and examination of whatever corroborating data are available. For example, it is widely recognized that occupational illnesses are poorly documented in the BLS Survey of Occupational Injuries and Illnesses, which captures only incident cases among active workers. Most illnesses that may have a relationship to work are not exclusively so related, and it is difficult for health practitioners to diagnose work-relatedness; few are adequately trained to make this assessment. Many of these illnesses have long latency and do not appear until years after people have left the employment in question. Surveillance programs may systematically undercount some categories of workers, such as contingent workers. Challenges posed by inadequate or inaccurate measurement systems should not drive programs out of difficult areas of study, and the ECs will need to be aware of such a possibility. In particular, contingent and informal working arrangements that place workers at greatest risk are also those on which surveillance information is almost totally lacking, so novel methods for measuring impact may be required.

In addition to measures of illness and injury, levels of exposure to chemical and physical agents and to safety and ergonomic hazards can be useful. Exposure or probability of exposure can serve as an appropriate proxy for disease or injury when a well-described occupational exposure-health association exists. In such instances, decreased exposure can be accepted as evidence that the end outcome of reduced illness has been achieved. That is particularly necessary in cases (such as exposure to asbestos) in which latency between exposure and disease outcome (lung cancer) makes effective evaluation of the relevant end outcome infeasible.

As an example of how exposure levels can serve as a proxy, the number of sites that exceed an OSHA Permissible Exposure Limit (PEL) or an American Conference of Governmental Industrial Hygienists threshold limit value is a quantitative measure of improvement of occupational health awareness and reduction of risk. In addition to exposure level, the number of people exposed and the distribution of exposure levels are important. Those data are available from multiple databases and studies of exposure. Apart from air monitoring, such measures of exposure as biohazard controls, reduction in requirements for use of personal protective equipment, and reduction of ergonomic risks are important.

Clearly, the commitment of industry, labor, and government to health and safety are critical external factors. Several measures of this commitment can be useful for the EC: monetary commitment of the groups, attitude, staffing, and surveys of relative level of importance. To the extent that the resources allocated to safety and health are limiting factors, the ECs should explicitly assess NIOSH performance in the context of constraints.

Questions to Guide the Evaluation Committee

1. What are the amounts and qualities of end-outcomes data (such as injuries, illness, exposure and productivity affected by health)?
2. What is the temporal trend in those data?
3. Is there objective evidence of improvements in occupational safety or health?
4. To what degree has the NIOSH program or subprogram been responsible for improvements in occupational safety or health?
5. If there is no time trend in the data, how do findings compare with data from other comparable US groups or the corresponding populations in other countries?
6. Is there evidence that external factors have affected outcome measures?

7. Has the program been responsible for outcomes outside the United States that have not been described in another category?

Assessment

For this part of the assessment, the EC should provide a qualitative assessment discussing the evidence of reductions in injuries and illnesses or their appropriate proxies (impacts).

III.B.9. Review of Other Outcomes

There may be health and safety impacts not yet appreciated, and other beneficial social, economic, and environmental outputs, including potential NIOSH impacts outside the United States. Many NIOSH study results and training programs may be judged to be important, or there may be evidence of implementation of NIOSH recommendations, outside the United States.

Questions to Guide the Evaluation Committee

1. Is the program likely to produce a favorable change that has not yet occurred or not been appreciated?
2. Has the program been responsible for other social, economic, security, or environmental outcomes?
3. Has the program's work had an impact on occupational health and safety in other countries?

Assessment

Evaluation by the EC may consist of a discussion of other outcomes, including positive changes that have not yet occurred; other social, economic, security, or environmental outcomes; and the impact that NIOSH has had on international occupational safety and health. It might also consider the incorporation of international research results into the NIOSH program of knowledge transfer for industry sectors.

III.B.10. Summary Evaluation Ratings and Rationale

An EC should use its expert judgment to rate the relevance and impact of the research program and its important subprograms by first summarizing its assessments of the subprograms and overall program according to the several items listed in Table 4. Table 4 is only a *worksheet* intended as an aid to the EC in its evalua-

TABLE 4 Evaluation Committee Worksheet to Assess Research Programs and Subprograms

Please respond to each with "major or important," "moderate," "likely," "limited," or "none."

Background Context for Program Impact

1.1 Evidence of reduction of risk factors in the workplace (intermediate outcome) and evidence that external factors affected reduction

1.2 Evidence of reduction in workplace exposure, illness, or injuries (end outcome) and evidence that external factors affected reduction

			Subprogram			
Addressing Charge 1	Activity Category	**Program**	1	...	...	n
1.3 Contributions of NIOSH research and transfer activities to changes in work-related practices	Research					
	Transfer					
1.4 Contributions of NIOSH research and transfer activities to reductions in workplace exposure, illness, or injuries	Research					
	Transfer					
1.5 Evidence of external factors preventing application of NIOSH research results	Research					
	Transfer					
1.6 Contribution of NIOSH research to enhancement of capacity in government or other research institutions	Research					
	Transfer					
1.7 Contributions of NIOSH research to productivity, security, or environmental quality (beneficial side effects)	Research					
	Transfer					
Addressing Charge 2						
2.1 Relevance of current and recently completed research and transfer activities to future improvements in workplace safety and health	Research					
	Transfer					
2.2 Progress in targeting research to areas of study most relevant to future improvements in occupational safety and health	Research					
	Transfer					

tion. Its purpose is to encourage the EC to summarize its work in one place and to concentrate on the subprograms and the items that will contribute to the final impact and relevance scores.

To set the context for this step in the evaluation of the impact of the research program in preparation to respond to charge 1, the EC will first need to consider the available evidence of changes in work-related risks and adverse effects and external factors related to the changes. That information should be organized as a prose response to items 1.1 and 1.2 in Table 4.

Next, the EC should review the responses to the questions in Sections III.B.6 through III.B.8 and systematically rate the impact of the research program and its subprograms by responding to items 1.3-1.7 in Table 4. To complete the table, the EC response should use one of the following five terms: "major or important," "moderate," "likely," "limited," or "none" (since 1995). The EC should evaluate separately the impact of the research and the impact of transfer activities. High ratings on items 1.3-1.7 require the committee's judgment that the program has contributed to outcomes. For example, outcomes have occurred earlier than they would have or are better than they would have been in the absence of the research program, or outcomes would have occurred in the absence of external factors beyond NIOSH's control or ability to plan around.

The EC should then assess the relevance of the research program and subprograms in preparation for addressing charge 2. The EC should review the responses to the questions in Sections III.B.2 through III.B.5 and rate the relevance of the research program and its subprograms by responding to items 2.1 and 2.2 in Table 4. The same five terms should be used ("major or important," "moderate," "likely," "limited," or "none") to evaluate separately the relevance of the research and the relevance of the transfer activities. Transfer activities occur in two contexts: (1) NIOSH efforts to translate intellectual products into practice and (2) efforts by stakeholders to take advantage of NIOSH products.

Final Program Ratings

To provide the final assessment of the research program for charge 1 (impact) and charge 2 (relevance), the ECs will use their expert judgment, their responses to the questions in Table 4, and any other appropriate information to arrive at one overall rating for the impact of the research program and one for its relevance to the improvement of occupational safety and health. In light of substantial differences among the types of research programs that will be reviewed and the challenge to arrive at a summative evaluation of both impact and relevance, however, the FC chose not to attempt to construct a single algorithm to produce the two final ratings.

Having completed Table 4, the EC should undertake its final assessment of the impact and relevance of the program. Final program ratings will consist of the numerical scores and prose descriptions of why the scores were given. As explained below, the ECs will summarize their responses to charges 1 and 2 by rating the relevance and impact of the NIOSH research program on five-point scales in which 1 is the lowest and 5 the highest rating. The FC has made an effort to establish mutually exclusive rating categories in the five-point rating scale; when the basis of a rating fits more than one category, the highest applicable score should be assigned. ECs will need to consider the impact and relevance of both NIOSH completed research and research in progress. In general, the assessment of impact will consider research completed, and the assessment of relevance will include research in progress related to likely future improvements. When assessing the relevance of the program, the EC should keep in mind how well the program has considered the frequency and severity of the problems being addressed, whether appropriate attention has been directed to both genders, vulnerable populations or hard-to-reach workplaces, and whether the different needs of large and small businesses have been accounted for.

The FC has some concern that the impact scoring system proposed below might be considered a promotion of the conventional occupational-health research paradigm that focuses on health-effect and technology research and not give much emphasis to socioeconomic and policy research and to surveillance and diffusion research (as opposed to activities) needed to effect change. Clearly, not all intermediate outcomes occur in the workplace. There are important outcomes much farther out on the causal chain that NIOSH can affect, and not all these can be defined as well-accepted intermediate outcomes. NIOSH, for example, has an important role to play in generating knowledge that may contribute to changing norms in the insurance industry, in health-care practice, in public-health practice, and in the community at large. The ECs may find that some of these issues need to be addressed and considered as important to influence the external factors that limit application of more traditional research findings. Given the rapidly changing nature of work and the workforce and some of the intractable problems in manufacturing, mining, and some other fields, the ECs are encouraged to think beyond the traditional paradigm.

Rating of Impact

5 = Research program has made a major contribution to worker health and safety on the basis of end outcomes or well-accepted intermediate outcomes.

4 = Research program has made a moderate contribution on the basis of end outcomes or well-accepted intermediate outcomes; research program generated important new knowledge and is engaged in transfer activities, but well-accepted intermediate outcomes or end outcomes have not been documented.
3 = Research program activities or outputs are going on and are likely to produce improvements in worker health and safety (with explanation of why not rated higher).
2 = Research program activities or outputs are going on and may result in new knowledge or technology, but only limited application is expected.
1 = Research activities and outputs are NOT likely to have any application.
NA = Impact cannot be assessed; program not mature enough.

Rating of Relevance

5 = Research is in highest-priority subject areas and highly relevant to improvements in workplace protection; research results in, and NIOSH is engaged in, transfer activities at a significant level (highest rating).
4 = Research is in high-priority subject area and adequately connected to improvements in workplace protection; research results in, and NIOSH is engaged in, transfer activities.
3 = Research focuses on lesser priorities and is loosely or only indirectly connected to workplace protection; NIOSH is not significantly involved in transfer activities.
2 = Research program is not well integrated or well focused on priorities and is not clearly connected to workplace protection and inadequately connected to transfer activities.
1 = Research in the research program is an ad hoc collection of projects, is not integrated into a program, and is not likely to improve workplace safety or health.

III.C. Identifying Significant Emerging Research (Addressing Charge 3)

Among the most challenging aspects of conducting research for the purpose of prevention of injury and illness is identifying new or emerging needs or trends and formulating an active research response that appropriately uses scarce resources in anticipation of those needs. Each EC should review the procedures that NIOSH has in place to identify needed research relevant to the NIOSH mission.

Each EC should review the success that NIOSH has had in identifying and addressing research to emerging issues. The review should include examination of leading indicators from appropriate federal agency sources, such as the Environmental Protection Agency, the Department of Labor, the National Institute of Standards and Technology, the National Institutes of Health, the Department of Defense, and the Department of Commerce. Those indicators should track new technologies, products, and processes and disease or injury trends.

One source of inputs deserving particular attention is the NIOSH HHE reports. NIOSH's HHE program is a separate legislatively mandated program that offers a potential mechanism to identify emerging research needs that could be incorporated as an input in each of the programs evaluated. The ECs should consider whether appropriate consideration has been given to findings from the HHE investigations as they are related to the research program under review.

Some additional indicators might include NIOSH and the NIOSH-funded FACE, the AOEC reports, the US Chemical Safety Board investigations, SENSOR and other state-based surveillance programs, and others. In addition, appropriate federal advisory committees and other stakeholder groups should be consulted to provide qualitative information.

The EC members should use their expert judgment both to evaluate what NIOSH has identified as emerging research targets (charge 2) and to respond to charge 3 by providing recommendations to NIOSH for additional research that NIOSH has not yet identified. An EC's response to charge 3 will consist primarily of recommendations for research in subjects that the EC considers important and of the committee's rationale.

Questions to Guide the Evaluation Committee

1. What information does NIOSH review to identify emerging research needs?
   a. What is the process for review?
   b. How often does the process take place?
   c. How are NIOSH staff scientists and NIOSH leadership engaged?
   d. What is the process for moving from ideas to formal planning and resource allocation?
2. How are stakeholders involved?
   a. What advisory or stakeholder groups are asked to identify emerging research targets?
   b. How often are such groups consulted, and how are suggestions followed up?
3. What new research targets have been identified for future development in the program under evaluation?
   a. How were they identified?

b. Were there lessons learned that could help to identify other emerging issues?
c. Does the EC agree with the issues identified and selected as significant and with the NIOSH response, or were important issues overlooked?
d. Is there evidence of unwise expenditure of resources on unimportant issues?

IV. EVALUATION COMMITTEE REPORT TEMPLATE

The following outline flows from the FC's review of the generalized logic model prepared by NIOSH, the request for information from NIOSH programs, and the assessment model described earlier in this report.

I. **Introduction**:
This section should be a brief descriptive summary of the history of the program (and subprograms) being evaluated, with respect to pre-NORA, NORA 1, and current and future plans of the research program presented by NIOSH. It presents the context for the research on safety and health; goals, objectives, and resources; groupings of subprograms; and any other significant or pertinent information. (A list of the NIOSH materials reviewed should be provided in an appendix to the EC report.)

II. **Evaluation of programs and subprograms (charges 1 and 2):**
A. Evaluation summary (includes a brief summary of the evaluation with respect to impact and relevance, scores for impact and relevance, and summary statements addressing charges 1 and 2).
B. Strategic goals and objectives: Describes assessment of the subprograms and overall program for relevance.
C. Review of inputs: Describes adequacy of inputs to achieve goals.
D. Review of activities: Describes assessment of the relevance and quality of the activities.
E. Review of research program outputs: Describes assessment of relevance, quality, and potential usefulness of the research program.
F. Review of intermediate outcomes and causal impact: Describes assessment of the intermediate outcomes and the causal attribu-

tion to NIOSH; includes the likely impacts and recent outcomes in the assessment.

G. Review of end outcomes: Describes the end outcomes related to health and safety and provides an assessment of the type and degree of causal attribution to NIOSH.

H. Review of other outcomes: Discusses other health and safety impacts that have not yet occurred; other beneficial social, economic, and environmental outcomes; and international dimensions and outcomes.

I. Summary of ratings and rationale (see Table 4).

III. Identification of needed research (charge 2):
The EC should assess the progress that the NIOSH program has made in targeting new research in the fields of occupational safety and health. There should be a discussion of the assessment process and results.

IV. Emerging research areas (charge 3):
The EC should assess whether the NIOSH program has identified significant emerging research areas that appear especially important in terms of their relevance to the mission of NIOSH. The EC should respond to NIOSH's perspective and add its own recommendations.

V. Recommendations for program improvement:
On the basis of the review and evaluation of the program, the EC may provide recommendations for improving the relevance of the NIOSH research program to health and safety conditions in the workplace and the impact of the research program on health and safety in the workplace as related to the research program under review.

Appendix A: List of the NIOSH and related materials collected in the process of the evaluation

V. FRAMEWORK COMMITTEE FINAL REPORT

At the conclusion of all individual program reviews, the FC will prepare a final report summarizing the findings of all the evaluating committees and providing NIOSH with an overall evaluation. All program ratings will be summarized and might be plotted graphically or with a Web chart.

The following is a proposed outline of the FC's final report:

I. Summary of national needs identified by the research programs reviewed.
   A. On the basis of the best available evidence, place those needs in the context of the overall estimated potential work-related disease and injury burden.
   B. Discuss the choices made and alternatives that might be the focus of current or future attention.
   C. Comment on programs not selected by NIOSH for evaluation by the National Academies.

II. Assessment of how well the program goals.
   A. Were matched to the research program needs.
   B. Were adjusted to new information and inputs as the field of interest changed or program results became available.

III. Assessment of NIOSH overall performance in the research programs reviewed.
   A. Distribution of available inputs.
   B. Activities and outputs.
   C. Intermediate outcomes.
   D. Summary assessment of significant differences among the programs.
   E. International impact.
   F. Leveraging of the NIOSH research activity with respect to other public and private research programs.
   G. Assessment of relative importance of external factors in permitting or preventing intermediate or end outcomes; attention paid to accounting for and planning within the constraints of external factors (not simply assigning lack of progress to external factors).

IV. Overall assessment of NIOSH impact on progress in reducing occupational injury and illness.
   A. Breakthrough knowledge.
   B. International impact.
   C. Addressing disparities.
   D. Targeting residual risks and intractable risks.
   E. Coordinating NIOSH research activity with respect to other public and private research programs.
   F. Impact on occupational safety and health.

V. Summary, Conclusions, and Recommendations.

B

Stakeholder Responses Online Survey

On February 28, 2007, the Committee to Review the NIOSH Respiratory Diseases Research Program sent out an invitation to stakeholders asking them to fill out an online questionnaire. A total of 25 people were sent an e-mail and/or letter. The letter and responses are given below (Box B-1).

RESPONSES TO ONLINE QUESTIONNAIRE

Question 1: Are you familiar with NIOSH activities and products related to occupational respiratory disease?

Respondent 1: Yes.

Respondent 2: Yes.

Respondent 3: Somewhat familiar, particularly regarding occupational infectious respiratory diseases.

Respondent 4: Yes, I know NIOSH approves respirators, dust masks, etc.

BOX B-1
Letter with Questionnaire

Hello,

The National Academies has been contracted by the National Institute of Occupational Science and Health (NIOSH) to conduct a review of its occupational respiratory disease activities.

As a part of this review the committee would like to give stakeholders an opportunity to comment on NIOSH's occupational respiratory disease activities and related products. The information that we receive from you will greatly help us in our review. Your responses are received anonymously, but the answers that are given will be made a part of public record.

We ask that you please take a moment to comment by clicking on the link below. http://www8.nationalacademies.org/survey/dels/niosh/niosh.htm

If you have any questions, please feel free to give myself or the staff officer on this study (K. John Holmes at 202-334-2045) a call or e-mail message. We would appreciate receiving responses by March 20th, but can accept responses up to May 1st. We appreciate your time in helping us with our task.

Best regards,
Jordan Crago

Question 2: What kind of experiences have you had working with the agency or its products?

Respondent 1: I am familiar with NIOSH's support of state-based surveillance activities for respiratory disease.

Respondent 2: Good except there could be more interaction between intramural staff and state surveillance programs with cooperative agreements to better target and use limited resources. Now only one person from NIOSH respiratory works with the states and other intramural people, and states do not coordinate activity even when they may be working on same specific respiratory issues

Respondent 3: As a CDC employee dealing with infection control matters for health care personnel, I have had to become somewhat familiar with respiratory protection for this occupational group. I have served on an informal CDC working group that included staff from NIOSH as well as others in CDC not at NIOSH. We discussed some research needs in the area of respiratory protection for health care personnel. This evolved into a group that organized a stakeholders' meeting in late 2005 to open discussion on a possible research agenda addressing respiratory protection for airborne infectious agents.

Respondent 4: I use to be an MSHA inspector and participated in some SCSR studies at Bruceton, PA and Morgantown, WV. Since returning to industry I have helped arrange a NIOSH visit to one of our mines.

Question 3: In what capacity have you interacted with NIOSH?

Respondent 1: I lead a state-based surveillance program. NIOSH supports our program with funding through a competitive grant process.

Respondent 2: Recipient of funds and reviewer of intramural and extramural activity.

Respondent 3: As a CDC colleague interested in similar issues.

Respondent 4: As an MSHA inspector (lab rat duties) and as a safety director for a coal company.

Question 4: Please comment on the relevance and impact of NIOSH's work over the past 10 years in any of the following areas:

a. Development, implementation, and evaluation of effective respiratory disease safety and health programs;
b. Evaluation of respiratory disease safety and health devices;
c. Development and use of engineering controls to reduce respiratory disease health hazards;
d. Improved understanding of occupational respiratory disease hazards and disorders through surveillance and investigation for risk factors.

Respondent 1: NIOSH has supported several states in the development of a surveillance programs for occupational asthma. Generally, these programs are able to identify new occupational asthmagens and identify some risk factors for OA. These programs are too restrictive and fail to include other occupational respiratory diseases, i.e., pneumoconiosis, or upper airway disorders. There is ineffective dissemination of information about new exposures and occupational respiratory disorders on a state level, i.e., diacetyl and brochiolitis obliterans. Failing the development of effective state partners in both the respiratory and other occupational diseases, there is little effective occupational public health at the state level.

Respondent 2: Identification of new occupational respiratory conditions has been major accomplishment. Work in surveillance is better than other NIOSH divisions with the periodic release of work-related lung disease surveillance report but more could be done particularly in coordination with the states. Development of respiratory controls have been excellent.

Respondent 3: NIOSH plays an important role in certification of respiratory protection equipment (RPE). In addition, they have conducted studies to assess the methods for evaluating the "fit" of RPE. Their strength is with agents for which there are known RELs, the difficulty is applying these principles to infectious agents from which RELs are not known nor for which measurement techniques currently exist. NIOSH staff have conducted studies on RPE "fit" that have had relevance to discussion of the need for annual fit testing of N95 respirators. They have conducted or sponsored studies to develop a more modern panel of subjects for fit testing of RPE.

Respondent 4: I know they've worked on mine rescue breathing apparatus studies and SCSR studies. I see their approval labels on the respirators we provide for our miners. That's about it.

Question 5: What do you see as and the major challenges and research needs over the past 10 years in occupational respiratory disease?

Respondent 1: 1) Identification of occupational respiratory disease clusters and subsequent investigation; 2) indoor air quality and mold exposures; and 3) the poor understanding of occupational respiratory by primary care and general physicians.

Respondent 2: To stay relevant by addressing major respiratory burdens and not just focusing on the less prevalent respiratory conditions unique to the work place.

Respondent 3: 1) Developing methods for detection and quantification of airborne infectious agents to be able to set exposure limits. 2) Development of standards for "fit" of RPE to which manufacturers should adhere. It is vital to develop RPE that have inherently good "fit" and for which fit testing may not be necessary. 3) Developing techniques to be able to determine if infectious agents are truly airborne, i.e., spread through droplet nuclei.

Respondent 4: I think the largest challenge was Congress doing away with the Bureau of Mines and only providing a small amount of funding to NIOSH to replace what was lost. I think more research needs to be done on respirable dust levels around continuous miners using flooded bed dust scrubbers.

Question 6: What do you see as significant emerging research needs in occupational respiratory disease?

Respondent 1: 1) Identification of new exposures for occupational asthma and respiratory disease, 2) effective delivery of quality medical care for occupational respiratory diseases (occupational health services research), and 3) effective education and controls for respiratory protection in emergency and disaster response.

Respondent 2: Continuing to determine the attributable risk for major respiratory diseases such as asthma, COPD, and lung cancer, and determining specific interventions and controls to reduce the incidence, morbidity, and mortality of these major respiratory conditions

Respondent 3: No answer

Respondent 4: I think more long-term studies using chest x-rays and pulmonary function tests need to be done to determine what today's miners and industrial workers are being exposed to. It will tell us how effective our current methods are.

Question 7: Do you have any other comments to offer about NIOSH research related to identification and control of occupational respiratory diseases?

Respondent 1: Get more state partners to improve surveillance and prevention programs!! Make the states work together and in partnership with the federal government.

Respondent 2: NIOSH has been precluded from using certain modalities such as specific antigen challenge testing which has hampered their research in the area of occupational asthma.

Respondent 3: NIOSH needs to continue to work collaboratively with other areas of CDC interested in these matters.

Respondent 4: No.

C

Bibliographic Information for the Committee to Review the NIOSH Respiratory Diseases Research Program

List of Documents and Other Items Submitted by the NIOSH Respiratory Diseases Research Program (RDRP) to the National Academies Evaluation Committee (EC) between October 10, 2006 and May 17, 2007. These documents are maintained on file at the National Research Council's (NRC) Board on Environmental Studies and Toxicology during the project and for a short time after its completion and, thereafter, are maintained in the NRC's archives.

TABLE C-1 List of Documents and Other Items Submitted by the NIOSH RDRP to the National Academies Evaluation Committee between October 10, 2006 and May 17, 2007

Document or Item	Brief Description	Date Submitted[a]
List of potential reviewers	RDRP's list of potential reviewers for inclusion by NA on EC	Spring 2006
Evidence package (EP)	Description of RDRP Activities, Outputs, Outcomes (13 chapters)	10.10.06
Electronic EP	Entire EP on "flash-drive" with hot links to Appendices (n = 458)	10.10.06
Slide presentations	PowerPoint format (163 slides) presentation of EP (selected parts)	10.27.06
Budgets 1996-2005[b]	Intramural and extramural budget information 1996-2005	11.21.06
Budget breakdowns[b]	Budget information broken down by research goals (Chapters 3-7 of EP)	11.21.06
Extramural projects 96-05[b]	Extramural research projects completed 1996-2005 plus budgets	11.21.06
List of HETAs 1994-2006[b]	Complete list of HETAs 1994-2006	11.21.06
Appendix A7-5[b]	List of RDRP nanotechnology publications (reports/abstracts/etc.)	11.21.06
Hard copies of appendices	Complete set of 458 appendices (supporting materials for the EP)	11.27.06
Slide presentation	PowerPoint format (22 slides) presentation of EP (Chapter 6)	12.06.06
Additional copies EP	Four additional copies EP hard copy and electronic versions	12.06.06
Word document	Infectious diseases (Weissman response to November 17, 2006 inquiry)	12.06.06
Word document	Future questions from Dr. Utell's letter[b]	12.06.06
Additional slide presentation (October 27, 2006)	Additional copies of slides presented on October 27, 2006	12.20.06
EP as Word document	Copy of EP in Word	12.20.06
A8-3	WoRLD Surveillance Report, hard copies to all EC members	01.06.07
Site visit agendas	Three agendas for site visits in Morgantown (January 16, 2007), Pittsburgh (January 17, 2007) and Cincinnati (January 18, 2007)	01.12.07

Document or Item	Brief Description	Date Submitted[a]
NIOSH 2007 Planning Guidance	Complete instructions for submission of research projects NIOSH-wide for the current fiscal years (includes sector-based plan)	01.31.07
89-128	NIOSH Planning document from 1986-1989	01.31.07
Response to Mark Utell's second request - Part 1[c]	Relationship between intra- and extramural programs; tabulation of outputs and outcomes; Asbestos Roadmap; Nano-budget	02.15.07
Response to Dr. Mark Utell's second request - Part 2[c]	Program Goals; nanotechnology issues; cancer program issues	03.01.07
Response to K. John Holmes e-mail request (February 16, 2007)	A5-5 (original and revised); tabulated contributions to IARC	03.02.07
Extramural funding	RDRP Extramural funding revision	03.16.07
External funding	External funding of intramural and extramural activities	03.19.07
List of RFAs	RFAs and funding 1996-2007	03.19.07
List of reimbursable items	Funds reimbursed to RDRP by other agencies	03.19.07
EP errata	Various corrections to EP	03.19.07
Budget revisions	Final budget revisions requested at NA meeting on March 22, 2007	04.07.07
Dr. Ray Sinclair's checklist[d]	Breakdown of requests channeled through Dr. Ray Sinclair	05.17.07
Inventory	Inventory of materials supplied to NAEC by (approximate) date Requested by K. John Holmes on May 15, 2007	05.17.07

[a]Dates are approximate.
[b]Response to Dr. Mark Utell's letter dated 11.02.2006.
[c]Response to Dr. Mark Utell's letter dated 12.06.2006.
[d]Detailed log of responses (largely broken out from Dr. Mark Utell's letters).

D

Biographic Information on the Committee to Review the NIOSH Respiratory Diseases Research Program

Mark J. Utell (Chair) is professor of medicine and environmental medicine, director of occupational and environmental medicine and former director of pulmonary and critical care medicine at the University of Rochester Medical Center. He serves as associate chairman of the Department of Environmental Medicine. His research interests have centered on the effects of environmental toxicants on the human respiratory tract. Dr. Utell has published extensively on the health effects of inhaled gases, particles and fibers in the workplace, indoor and outdoor environments. He is the coprincipal investigator of an EPA Particulate Matter Center and chair of the Health Effects Institute's Research Committee. He has served as chair of EPA's Environmental Health Committee and on the Executive Committee of the EPA Science Advisory Board. He is a former recipient of the NIEHS Academic Award in Environmental and Occupational Medicine. Dr. Utell is currently a member of the National Research Council's (NRC) Board on Environmental Studies and Toxicology. He previously served on the NRC Committee on Research Priorities for Airborne Particulate Matter, the Institute of Medicine's (IOM) Committee to Review the Health Consequences of Service during the Persian Gulf War; and the IOM Committee on Biodefense Analysis and Counter-measures. He received his M.D. from Tufts University School of Medicine in 1972.

John R. Balmes is professor of medicine at the University of California, San Francisco, and chief of the Division of Occupational and Environmental Medicine at San Francisco General Hospital. He is also professor of Environmental Health

Sciences at the University of California, Berkeley, and director of the Northern California Center for Occupational and Environmental Health. His research is in the area of occupational and environmental respiratory disease. He studies the acute effects of inhalation exposures to ambient air pollutants in his human exposure laboratory at San Francisco General Hospital and the chronic effects of such exposures in epidemiological studies with collaborators at the University of California, San Francisco, and the University of California, Berkeley. He is also interested in genetic determinants of responses to air pollutants. For the past two years, Dr. Balmes has been leading research, funded by the CDC, to assist in the development of a national program to link environmental hazards with health outcome data to improve the tracking of diseases potentially related to environmental exposures. Dr. Balmes received the Environmental and Occupational Medicine Academic Award from NIEHS,1991-1996. Dr. Balmes received his M.D. from Mt. Sinai School of Medicine in 1976.

Paul D. Blanc is professor of medicine, Endowed Chair and Chief of the Division of Occupational and Environmental Medicine at the University of California School of Medicine, San Francisco. Dr. Blanc also serves as associate medical director of the California Poison Control System, San Francisco Division. Dr. Blanc is board certified in occupational medicine and internal medicine and holds a certificate in medical toxicology. His research interests are in the areas of epidemiology of occupational lung disease, asthma outcomes, and occupational toxicology. He was a member of the IOM Committee on Poison Prevention and Control. He received his M.D. from the Albert Einstein College of Medicine. Dr. Blanc is the author of *How Everyday Products Make People Sick* (UC Press).

Elizabeth S. Chamberlin is vice president for safety and training at Massey Energy. Before that she was the general manager for safety for CONSOL Energy Incorporated. She was formerly part of the General Counsel's office for CONSOL Energy. She is trained as a mining engineer and has experience as a miner, operating engineer, and assistant foreman while at the United States Steel Mining Company. She chairs the National Mining Association's Health and Safety Sub-Committee. She received her J.D. from Duquesne School of Law and an M.B.A. from Waynesburg College.

Rogene F. Henderson is a senior biochemist and toxicologist emeritus in the Experimental Toxicology Program of the Lovelace Respiratory Research Institute. She is also a clinical professor in the College of Pharmacy at the University of New Mexico in Albuquerque. Her major research interests are in the use of bronchoalveolar lavage fluid analyses to detect and characterize biomarkers of developing lung disease, the toxicokinetics of inhaled vapors and gases, and the use of biological markers of exposure and of effects to link environmental exposure to disease. She has served on a number of scientific advisory boards, including those of DOE, EPA, NIEHS, and the U.S. Army. She was recently appointed chair of EPA's Clean Air Scientific Advisory Committee. Dr. Henderson is a National Associate of the National Academies, and is a former member of the Board on Environmental Studies and Toxicology. She has served on numerous NRC committees, and is currently serving on the Committee on Human Health Risks from Trichloroethylene and the Committee on Asbestos: Selected Health Effects. She received her Ph.D. in chemistry from the University of Texas in 1960.

David M. Mannino is associate professor of medicine and a clinician and scientist in the Division of Pulmonary and Critical Care Medicine at the University of Kentucky Medical Center. Dr. Mannino specializes in chronic bronchitis and emphysema, asthma, and health-effects related to tobacco smoke. His current research projects include health effects related to tobacco smoke, metals and the lungs, and outcomes and comorbidities of chronic bronchitis and emphysema. Dr. Mannino received the Soffer Research Award in 2003 from the American College of Chest Physicians and was elected Advocate of the Year by the American Lung Association in 2003 for work on a smoking ordinance in DeKalb County, Georgia. Dr. Mannino earned an M.D. from the Jefferson Medical College in 1981.

James A. Merchant is professor of occupational and environmental health and dean of the College of Public Health at the University of Iowa. Dr. Merchant's expertise is in medicine, public-health, industrial health, federal government agency administration, rural health policy, preventive medicine, occupational medicine, environmental health, and epidemiology. His research interests include occupational and environmental lung disease, rural health outcomes, rural health care delivery, public health policy, environmental health sciences, and international health. Dr. Merchant is a member of IOM and he serves on the Roundtable on Environmental Health Sciences, Research, and Medicine. He was honored with the James P. Keogh Award by the National Institute for Occupational Safety and Health/Centers for Disease Control and Prevention in 2003. Dr. Merchant received his M.D. from the University of Iowa and a Ph.D. in epidemiology from the University of North Carolina.

Jacqueline Nowell is the director of the Occupational Safety and Health Office at the United Food and Commercial Workers International Union. Ms. Nowell and her staff develop and monitor ergonomic programs in the red meat, poultry, and retail industries. They develop educational materials and conduct training programs for local union stewards and leadership on a variety of safety and health issues in the union's represented industries. She is a member of the American Public Health Association and American Industrial Hygiene Association as well as serving on National Institute for Occupational Safety and Health/National Occupational Research Agenda Traumatic Injuries and Special Populations at Risk Teams. She is currently a board member on the District of Columbia Occupational Safety and Health Board that establishes policies related to occupational safety and health issues in the District of Columbia. Ms. Nowell received her Master's in Public Health from the University of California, Los Angeles and is a certified industrial hygienist. She has worked for the New York Committee for Occupational Safety and Health and was an assistant professor at Hunter College's School of Health Sciences/Environmental and Occupational Health Science Program.

Charles Poole is associate professor in the Department of Epidemiology at the University of North Carolina School of Public Health. Previously, he was with the Boston University School of Public Health. Dr. Poole's work focuses on the development and utilization of epidemiologic methods and principles, including problem definition, study design, data collection, statistical analysis, and interpretation and application of research results, including systematic review and meta-analysis. His research experience includes studies in environmental and occupational epidemiology and other substantive areas. Dr. Poole was an epidemiologist in the Office of Pesticides and Toxic Substances of the U.S. Environmental Protection Agency for 5 years and worked for a decade as an epidemiologic consultant, both with a firm and independently. Dr. Poole was a member of the IOM Committee on Gulf War and Health: Review of the Literature on Pesticides and Solvents and the NRC Committees on Estimating the Health-Risk-Reduction Benefits of Proposed Air Pollution Regulations and Fluoride in Drinking Water. He received his Sc.D. in epidemiology from the Harvard School of Public Health in 1989.

Richard B. Schlesinger is associate dean for academic affairs and research of Dyson College of Arts and Sciences and professor and chair of the Department of Biology-Health Sciences at Pace University, New York. He was previously director of the toxicology program at the New York University School of Medicine. In 2006, he received the Herbert E. Stokinger award from the American Conference of Industrial Hygienists for his contributions to the field of environmental toxicology. He has served on several NRC committees includi ng the Pulmonary Toxicology and Research Priorities for Airborne Particulate Matter and is currently a member of the Subcommittee on Acute Exposure Guideline Levels. He is an associate editor of the journal *Inhalation Toxicology*. He received his Ph.D. in biology and environmental health from New York University.

Noah S. Seixas is the Rohm and Hass Professor of Environmental and Occupational Health Sciences in the School of Public Health and Community Medicine at the University of Washington, and the director of the Northwest Center for Occupational Health and Safety. His research interests and expertise are in exposure assessments in the context of retrospective, cross-sectional, and prospective epidemiology studies, and have included studies on silica exposure, irritant exposures in aluminum smelting, and organic dust exposures. Over the past eight years, Dr. Seixas has focused largely on noise exposure in the construction industry. In addition to studies in construction, Dr. Seixas is currently working on risks in "precarious employment," including risks faced by day laborers. Dr. Seixas is an assistant editor of the *Annals of Occupational Hygiene* and serves on the editorial boards of the *Journal of Occupational and Environmental Hygiene* and the *American Journal of Industrial Medicine.* He received his M.S. from the Harvard School of Public Health in 1984 and his Ph.D. in industrial health from the University of Michigan in 1990.

Ira B. Tager is professor of epidemiology in the Division of Public Health, Biology, and Epidemiology at the University of California, Berkeley, and is codirector and principal investigator for the Center for Family and Community Health. Dr. Tager's research interests include, among others, the development of exposure assessment instruments for studies of health effects of chronic ambient ozone exposure in childhood and adolescence, effects of ozone exposure on pulmonary function, and the effects of oxidant and particulate air pollution on cardiorespiratory morbidity and mortality as well as air-pollution-related morbidity from asthma in children. Dr. Tager was a member of the NRC committee on Air Quality in Passenger Cabins of Commercial Aircraft and currently serves on the committee on the Effects of Changes in New Source Review Programs for Stationary Air Pollutants. He also serves as a member of the Research Committee for the Health Effects Institute. Dr. Tager received an M.D. from the University of Rochester School of Medicine and a M.P.H. from the Harvard School of Public Health.

David H. Wegman is dean of the School of Health and Environment at the University of Massachusetts, Lowell. He also serves as adjunct professor at the Harvard School of Public Health. Dr. Wegman's research involves epidemiologic studies of occupational respiratory disease, musculoskeletal disorders, and cancer. Recent work has focused on the examination of health and safety risks among construction workers involved in the building of the Third Harbor Tunnel and the underground Central Artery in Boston, and the study of the relationship of work risks and age both among child laborers and older adults. He has also written on public health and policy issues concerning hazard and health surveillance, methods of exposure assessment for epidemiologic studies, the development of alternatives to regulation and the use of participatory methods to study occupational health risks. Dr. Wegman served as chair of the NRC-IOM Committee on Health and Safety Needs of Older Workers and the Committee on the Health and Safety Consequences of Child Labor. He has also been a member of the NRC-IOM Panel on Musculoskeletal Disorders and Work, the IOM Committees to Review the Health Consequences of Service During the Persian Gulf War and to Review Gender Differences in Susceptibility to Environmental Factors, and he currently chairs the Committee on the Review of NIOSH Research Programs, which serves as the Framework Committee for this and other specific NIOSH reviews. He received his M.D. from Harvard Medical School.